ELSEVIER'S DICTIONARY OF ENVIRONMENT

in
English, French, Spanish and Arabic

compiled by

M. BAKR
Nyon, Switserland

1998
ELSEVIER
Amsterdam – Lausanne – New York – Oxford – Shannon – Singapore – Tokyo

ELSEVIER SCIENCE B.V.
Sara Burgerhartstraat 25
P.O. Box 211, 1000 AE Amsterdam, The Netherlands

First edition 1998

Library of Congress Cataloging in Publication Data
A catalog record from the Library of Congress has been applied for.

ISBN: 0-444-82966-0

♾ The paper used in this publication meets the requirements of ANSI/NISO Z39.48-1992 (Permanence of Paper).
Printed in The Netherlands.

Preface

Environmental concerns are important items on the political and economic agendas of most countries, whether developed or developing. In a world conscious of its environmental problems, studies in this domain have become an established discipline. Every new branch of science needs to define relevant terms to be used through the medium of language. Hence, terminologies are developed.

The subject of environment is global because it touches everyone. Individuals, specialists or institutions concerned with the health of our planet need to mean the same thing when communicating with each other, notwithstanding language differences. The compilation of a multilingual dictionary on the environment is not easy. The inclusion or omission of certain terms is an important question. However, the choice of terms, in this text, has been made on two assumptions: (1) the most commonly used terms, (2) their translatability into other languages without losing their intended meaning. Consideration of regional differences of the same language has been considered.

The main objective of this work is to assist those involved in environmental activities in their attempt to make the world cleaner and to sustain its natural resources for future generations. This, I hope, will help to produce documentation for meetings and conferences as well as for drafting papers for training, information and communication purposes.

I would like to thank those who helped me in producing this dictionary, particularly, Charles Serex, a chemical engineer, who has extensive scientific knowledge in this field and linguistic sensitivity. I am grateful also to Huguette for her accurate proofreading. My thanks go to Sherif and his team for typing and preparing the manuscript.

A final word must be said: I do not claim that this work is perfect because certain equivalents may be open to challenge. For this, I beg the user's forbearance.

M. Bakr
May 1998
Nyon, Switzerland

The subfields of the dictionary are:

air pollution
biological diversity
biomass energy
biosafety
biotechnology
climate change
coastal environment
deforestation
desertification
endangered species
energy conservation
environmental economics
forest conservation
freshwater pollution
global warming
greenhouse gases
human settlements
living marine sources
marine environment
mountain ecosystem
ozone layer
resources management
soil degradation
sustainable development
tropical ecosystem

Basic Table

A

1 **abandoned agricultural lands**
f terres agricoles à l'abandon
e tierras agrícolas abandonadas
ارض زراعية مهملة

2 **abatement**
f réduction
e reducción
الحد من

3 **abiotic**
f abiotique
e abiótico
لا احيائي

4 **abiotic stress**
f stress abiotique
e presión abiótica
اجهاد لا احيائي

5 **abrasive**
f abrasif
e abrasivo
كاشط

6 **absorber**
f absorbant
e absorbente
ماص

7 **absorber vessel**
f cuve d'absorption
e cuba de absorción
وعاء الامتصاص

8 **absorbing ability**
f pouvoir absorbant d' un sol
e poder de absorción
قدرة المص

9 **absorptive complex**
f complexe absorbant
e complejo absorbente
مجمع امتصاصي

10 **abundance**
f abondance
e abundancia
وفرة

11 **acceptable daily intake**
f dose journalière admissible
e dosis diaria aceptable
الجرعة اليومية المقبولة

12 **acceptance test**
f essai d'homologation
e prueba de aceptación
اختبار قبول

13 **access arrangement**
f mécanisme d' accès
e mecanismo de acceso
ترتيب الوصول

14 **access roads**
f routes de desserte
e carreteras de acceso
طرق الوصول

15 **access to the sea**
f accès à la mer
e acceso al mar
الوصول إلى البحر

16 accident management
f gestion des accidents
e medidas para hacer frente a los accidentes
ادارة الحوادث

17 accidental release
f introduction accidentelle
e escape al medio ambiente
اطلاق عرضي

18 accidental release of organisms
f introduction accidentelle de microorganismes (dans l'environnement)
e escape accidental de microorganismos(en el medio ambiente)
إطلاق عرضى للكائنات الحية

19 accidental spill
f déversement accidentel
e derrame (accidental)
انسكاب عرضى

20 accumulation in body tissues
f accumulation dans les tissus organiques
e acumulación en los tejidos del cuerpo
تراكم فى أنسجة الجسم

21 acid equivalents
f équivalents acides
e equivalentes ácidos
مكافئات حمضية

22 acid fallout
f retombées acides
e precipitación ácida
تساقط حمضى

23 acid fog
f brouillard acide
e niebla ácida
ضباب حمضي

24 acid mist
f brume acide
e neblina ácida
رذاذ حمضي

25 acid rain
f pluie acide
e lluvia ácida
امطار حمضية

26 acid(ic) deposit(ion)
f retombée acide
e deposición ácida
ترسب حمضي

27 acid(ic) precipitation
f précipitation (s) acide (s)
e precipitación ácida
تهطل حمضي

28 acid(ic) snow
f neige acide
e nieve ácida
ثلج حمضي

29 acid-affected
f atteint par la pollution acide
e contaminado por el ácido
متأثر بالاحماض

30 acid-causing substance
f substance acidifiante
e sustancia acidificante
مادة مولدة للحمض

31 **acid-resistant organism**
f organisme résistant aux acides
e organismo resistente a los ácidos
كائن مقاوم للاحماض

32 **acidification**
f acidification
e acidificación
تحمض

33 **acidifying substance**
f substance acidifiante
e sustancia acidificante
مادة محمضة

34 **acids**
f acides
e ácidos
أحماض

35 **acoustic insulation**
f isolation acoustique
e aislamiento acústico
عزل صوتى

36 **acoustics**
f acoustique
e acústica
صوتيات

37 **activated carbon**
f charbon actif
e carbón activado
فحم منشط

38 **activated sludge**
f boues activées
e fangos activados
حمأة منشطة

39 **activation of a response plan**
f déclenchement d' un plan d' intervention
e puesta en marcha de un plan de intervención
بدء تنفيذ خطة استجابة

40 **active**
f actif
e activo
نشط

41 **acutely toxic pesticide**
f pesticide à toxicité aiguë
e plaguicida a toxicidad aguda
مبيد آفات حاد السمية

42 **acyclic**
f acyclique
e acíclico
لا حلقي

43 **adaptable species**
f espèces adaptables
e especies adaptables
أنواع متكيفة

44 **add-on hardware**
f dispositif additionnel
e equipo suplementario
معدات اضافية

45 **add-on technology**
f techniques de rattrappage
e tecnología suplementaria
تكنولوجيا الاضافة

46 **add-on waste treatment**
f traitement des déchets par des dispositifs auxiliaires
e tratamiento suplementario de desechos
معالجة الفضلات بمعدات اضافية

47 **adiabatic process**
f processus adiabatique
e proceso adiabático
عملية بلا تبادل حراري

48 **adsorbent**
f adsorbant
e adsorbente
ماز

49 **adsorption**
f adsorption
e adsorción
امتزاز

50 **advance planting**
f plantation anticipée
e plantación anticipada
زرع مسبق

51 **adverse climate change**
f évolution nuisible du climat
e cambios climáticos negativos
تغير مناخي ضار

52 **adverse effect**
f effet nocif
e efecto negativo
اثر ضار

53 **adverse environmental effect**
f effet nocif pour l' environnement
e efecto perjudicial para el medio ambiente
اثر بيئي ضار

54 **adverse health effect**
f effet nocif pour la santé
e efecto nocivo para la salud
اثر ضار بالصحة

55 **adverse impact**
f incidences négatives
e repercusiones negativas
تأثير ضار

56 **aerial plankton**
f plancton atmosphérique
e plancton atmosférico
عوالق هوائية

57 **aerobic conditions**
f conditions aérobiques
e estados aerobios
ظروف هوائية

58 **aerobic processes**
f processus aérobies
e procesos aeróbicos
عمليات هوائية

59 **aeronautical climatology**
f climatologie aéronautique
e climatología aeronáutica
علم مناخ الملاحة الجوية

60 **aerosol**
f aérosol
e aerosol
هباء جوي

61 aerosol (spray-) can
f bombe aérosol
e atomizador de aerosoles
مرذاذ

62 aerosol dispenser
f générateur d'aérosol
e atomizador de aerosoles
مرذاذ

63 aerosol load
f teneur en aérosols
e carga de aerosoles
كمية الهباء الجوى

64 aerosols
f aérosols
e aerosoles
ايروصولات

65 affected country
f pays touché
e país afectado
بلد متأثر

66 affordable water
f eau à des prix raisonnables
e agua a precios razonables
مياه رخيصة

67 afforestation
f boisement
e forestación
تشجير

68 aflatoxin
f aflatoxine
e aflatoxina
سم الفطريات

69 African horse sickness
f peste équine
e peste equina
طاعون الخيل الافريقي

70 after-care disposal sites
f surveillance des sites d'élimination
e mantenimiento de vertederos
العناية اللاحقة بمواقع التصريف

71 aftershock
f réplique (d'un séisme)
e temblor secundario
هزة لاحقة

72 age-class system
f méthode des classes d' âge
e método por clases de edad
نظام التصنيف العمري

73 age-old forest
f forêt centenaire
e bosque secular
غابة معمرة

74 Agenda 21
f Action 21
e Programa 21
جدول اعمال القرن ٢١

75 aggradation
f exhaussement
e agradación
إطماء

76 aggregator
f coefficient d'agrégation
e coeficiente de agregación
مجمّع

77 agricultural biotechnologies
f biotechnologies agricoles
e biotecnologías agrícolas
تكنولوجيا حيوية زراعية

78 agricultural ecology
f écologie agricole
e ecología agrícola
ايكولوجيا زراعية

79 agricultural economics
f économie agricole
e economía agrícola
اقتصاد زراعى

80 agricultural engineering
f génie agricole
e ingeniería agrícola
هندسة زراعية

81 agricultural equipment
f équipement agricole
e maquinaria agrícola
معدات زراعية

82 agricultural land
f terres agricoles
e terrenos agrícolas
أرض زراعية

83 agricultural legislation
f législation agricole
e legislación agrícola
تشريع زراعى

84 agricultural management
f gestion agricole
e gestión agrícola
إدارة زراعية

85 agricultural methods
f méthodes agricoles
e métodos agrícolas
وسائل زراعية

86 agricultural pests
f parasites agricoles
e plagas agrícolas
آفات زراعية

87 agricultural practices
f pratiques agricoles
e prácticas agrícolas
ممارسات زراعية

88 agricultural production
f production agricole
e producción agrícola
إنتاج زراعى

89 agricultural run-off
f ruissellement des terres agricoles
e escurrimiento de tierras agrícolas
صرف زراعي

90 agricultural storage
f stockage des produits agricoles
e almacenamiento de productos agrícolas
تخزين زراعى

91 agricultural waste site
f décharge de déchets agricoles
e vertedero de desechos agrícolas
موقع نفايات زراعية

92 agricultural wastes
f déchets agricoles
e desechos agrícolas
نفايات زراعية

93 agriculture
f agriculture
e agricultura
الزراعة

94 agro-ecological zone
f zone agro-écologique
e zona agroecológica
منطقة زراعية ايكولوجية

95 agro-fish processing industry
f industrie de transfomation des produits halieutiques
e industria de elaboración de productos agropiscícolas
صناعة تجهيز المنتجات السمكية

96 agro-food (stuffs) industry
f industrie agroalimentaire
e industria agroalimentaria
صناعة الاغذية الزراعية

97 agro-industry
f agro-industrie
e agroindustria
صناعة زراعية

98 agro-pastoralist
f agro-pasteur
e productor agropecuario
مزارع - راع

99 agro-sylvo-pastoral system
f système agro-sylvo-pastoral
e sistema agro-silvo-pastoral
نظام زراعي - حرجي - رعوي

100 agrochemicals
f produits agrochimiques
e productos agroquímicos
كيماويات زراعية

101 agroclimatic
f agroclimatique
e agroclimático
مناخي زراعي

102 agroclimatological zoning
f zonage agroclimatique
e zonación agroclimática
تقسيم المناطق حسب علم المناخ الزراعي

103 agroclimatology
f agroclimatologie
e agroclimatología
علم المناخ الزراعي

104 agroforestry
f agroforesterie
e agrosilvicultura
حراجة زراعية

105 agrometeorology
f météorologie agricole
e meteorología agrícola
الارصاد الجوية الزراعية

106 air barrier
f rideau de bulles
e cortina de burbujas
حاجز هوائي

107 air chemistry
f chimie de l'atmosphère
e química de la atmósfera
كيمياء الهواء

108 air column
f colonne atmosphérique
e columna de aire
عمود الهواء

109 air conditioning
f climatisation
e climatización
تكييف الهواء

110 air craft
f avions
e aeronaves
طائرة

111 air current
f courant aérien
e corriente de aire
تيار هوائى

112 air intrusion
f pénétration d'air
e intrusión de aire
اقتحام هوائي

113 air mass
f masse d'air
e masa de aire
كتلة هوائية

114 air monitoring
f surveillance de l'air
e vigilancia de la contaminación atmosférica
رصد الهواء

115 air parcel
f particule dans l'air
e partícula en el aire
قطعة هوائية

116 air pollutant
f polluant atmosphérique
e contaminante atmosférico
ملوث جوي

117 air pollution
f pollution atmosphérique
e contaminación atmosférica
تلوث الهواء

118 air pollution control
f lutte contre la pollution atmosphérique
e lucha contra la contaminación atmosférica
مكافحة تلوث الهواء

119 air pollution load
f charge de polluants atmosphériques
e carga de contaminantes atmosféricos
كمية الملوثات الجوية

120 air quality
f qualité de l'air
e calidad del aire
نوعية الهواء

121 air quality management
f gestion de la qualité de l'air
e gestión de la calidad del aire
إدارة نوعية الهواء

122 air surveillance
f surveillance réglementaire de la pollution de l'air
e vigilancia de la calidad del aire
رصد الجو

123 air traffic regulations
f réglementation du trafic aérien
e reglamento del tráfico aéreo
لوائح النقل الجوى

124 air transportation
f transports aériens
e transporte aéreo
نقل جوى

125 air-fuel ratio
f mélange air-carburant
e mezcla de aire y carburante
نسبة الهواء الى الوقود

126 air-sea boundary layer
f couche limite entre l'atmosphère et l'océan
e capa de contacto entre la atmósfera y el océano
الطبقة الحدية بين الهواء والبحر

127 air-sea interaction
f interaction air-mer
e interacción aire-mar
التفاعل بين الهواء والبحر

128 air-water interaction
f interaction air-eau
e interacción aire-agua
تفاعل بين الهواء والماء

129 airborne contaminant
f contaminant atmosphérique
e contaminante transportado por el aire
ملوث جوي

130 airborne emissions
f émissions dans l'atmosphère
e emisiones transportadas por el aire
انبعاثات جوية

131 airborne fraction of carbon
f fraction du carbone transporté par l'air
e fracción del carbono transportada por el aire
الاجزاء الكربونية العالقة بالهواء

132 airborne particulates
f particules en suspension dans l'air
e partículas en suspensión en el aire
جسيمات عالقة بالهواء

133 airborne pollutant concentration
f teneur de l'air en polluants
e concentración de contaminantes transportados por el aire
تركز الملوثات العالقة بالهواء

134 airborne pollution
f pollution atmosphérique
e contaminación transportada por el aire
تلوث جوي

135 airborne transport
f transport atmosphérique
e transporte atmosférico
النقل بواسطة الجو

136 aircraft engine emissions
f émission des moteurs d'avion
e emisiones de motores de aeronaves
انبعاثات محركات الطائرات

137 aircraft noise
f bruits d'avions
e ruido de aeronaves
ضوضاء الطائرات

138 airflow
f écoulement de l' air
e flujo de aire
تدفق هوائي

139 airports
f aéroports
e aeropuertos
مطارات

140 albedo
f albédo
e albedo
البياض

141 alcohol fuel
f alcool carburant
e carburante de alcohol
وقود كحولي

142 algae
f algues
e algas
طحالب

143 algae-bearing layer
f banc d' algues
e capa de algas
طبقة حاملة للطالحب

144 algae bloom
f florissement des algues
e lozanía de algas
تكاثر الصحالب

145 algae control
f lutte contre la prolifération des algues
e lucha contra la proliferación de las algas
مكافحة الطحالب

146 algal biomass
f biomasse algale
e biomasa de algas
كتلة حيوية طحلبية

147 algal bloom
f efflorescence algale
e floración de algas
تكاثر

148 algal layer
f masse algale
e capa de algas
طبقة طحلبية

149 alicyclic
f alicyclique
e alicíclico
دهني حلقي

150 aliphatic compound
f hydrocarbure aliphatique
e compuesto alifático
مركب دهني

151 alkali fumes
f vapeurs alcalines
e vapores alcalinos
ابخرة قلوية

152 alkali lands
f terres alcalines
e tierras alcalinas
أرض قلوية

153 alkali metal
f métal alcalin
e metal alcalino
فلز قلوي

154 alkaline - earth metal
f métal alcalino - terreux
e metal alcalinotérreo
فلز ترابي قلوي

155 alkaline buffer
f tampon alcalin
e amortiguador alcalino
حاجز قلوي

156 alkalinization
f alcalinisation
e alcalinización
المعالجة بمادة قلوية

157 allergenicity
f propriétés allergisantes
e alergenicidad
خاصية اثارة الحساسية

158 allergens
f allergènes
e alergenos
مثيرات الحساسية

159 alleviating measures
f mesures d' atténuation
e medidas de mitigación
تدابير مخففة

160 alleviating of poverty
f atténuation de la pauvreté
e mitigación de la pobreza
تخفيف حدة الفقر

161 allied species
f espèces apparentées
e especies similares
الانواع المتقاربة

162 alluviation
f atterrissement
e aterramiento
ترسب الطمي

163 alternative development
f autre développement
e otra modalidad de desarrollo
تنمية بديلة

164 alternative fuel
f carburant de remplacement
e carburante alternativo
وقود بديل

165 alternative product
f produit de remplacement
e producto alternativo
منتج بديل

166 alternative strategy
f stratégie de rechange
e estrategia alternativa
استراتيجية بديلة

167 alternative technology
f technique alternative
e tecnología alternativa
تكنولوجيا بديلة

168 altitude distribution
f répartition verticale
e distribución vertical
توزع عمودي

169 alumina
f alumine
e alúmina
أكسيد الألمنيوم

170 aluminum industry
f industrie de l'aluminium
e industria del aluminio
صناعة الألومنيوم

171 ambient air
f air ambiant
e aire ambiente
الهواء المحيط

172 ambient air quality standard
f norme de qualité de l'air ambiant
e norma de calidad del aire ambiente
معيار نوعية الهواء المحيط

173 amine ring
f cycle amine
e anillo de amina
حلقة نشادرية

174 ammonia
f ammoniac
e amoníaco
غاز النشادر

175 ammonolysis
f ammonolyse
e amonolisis
انحلال نشادري

176 ammoxidation
f ammoxydation
e amoxidación
اكسدة نشادرية

177 amount of precipitation
f niveau de précipitation
e nivel de la precipitación
كمية الامطار

178 amphibians
f amphibiens
e anfibios
برمائيات

179 amplification factor
f coefficient d' incidence
e factor de amplificación
عامل التضخيم

180 anabatic wind
f vent anabatique
e viento anabático
ريح صاعدة

181 anaerobic conditions
f conditions anaérobiques
e condiciones anaerobias
ظروف لاهوائية

182 anaerobic processes
f processus anaérobies
e procesos anaeróbicos
عمليات لا هوائية

183 analytical chemistry
f chimie analytique
e química analítica
كيمياء تحليلية

184 analytical equipment
f appareils d'analyse
e equipo analítico
معدات تحليلية

185 animal behaviour
f comportement animal
e comportamiento animal
سلوك الحيوان

186 animal density
f densité du cheptel
e densidad de la población animal
الكثافة الحيوانية

187 animal diseases
f maladies animales
e enfermedades de animales
أمراض الحيوان

188 animal dung as fuel
f bouse comme combustible
e estiércol animal usado como combustible
روث الحيوانات كوقود

189 animal ecology
f écologie animale
e ecología animal
ايكولوجيا الحيوان

190 animal genetic resources for agriculture
f ressources zoogénétiques pour l'agriculture
e recursos zoogenéticos para la agricultura
الموارد الجينية الحيوانية لاغراض الزراعة

191 animal genetics
f génétique animale
e genética animal
جينات حيوانية

192 animal husbandry
f zootechnie
e zootecnia
رعاية الحيوان

193 animal nutrition
f nutrition animale
e nutrición animal
تغذية الحيوان

194 animal physiology
f physiologie animale
e fisiología animal
علم وظائف اعضاء الحيوان

195 animal products
f produits animaux
e productos animales
منتجات حيوانية

196 animal resources
f ressources animales
e recursos animales
موارد الحيوان

197 anomalous fluctuations
f fluctuations irrégulières
e fluctuaciones anormales
تقلبات شاذة

198 anoxia
f anoxie
e anoxia
نقص الاكسجين

199 antarctic ecosystems
f écosystèmes antarctiques
e ecosistemas antárticos
النظم الايكولوجية لأنتاركتيكا

200 antarctic region
f régions antarctiques
e región antártica
منطقة أنتاركتيكا

201 anthropogenic
f anthropique
e antropógeno
من صنع الانسان

202 anti-desertification programme
f programme de lutte contre la désertification
e programa de lucha contra la desertificación
برنامج مكافحة التصحر

203 anti-feedant protein
f protéine antinutritionnelle
e proteína antialimentaria
بروتين مانع للمغذيات

204 anti-poverty programme
f programme de lutte contre la pauvreté
e programa de lucha contra la pobreza
برنامج مكافحة الفقر

205 anti-trade
f contre - alizé
e viento contraalisio
الرياح المضادة للريح التجارية

206 anticipatory approach
f anticipation
e criterio previsor
نهج ترقبي

207 antioxidant
f antioxydant
e antioxidante
مقاوم للتأكسد

208 antipollution incentives
f encouragement à l'anti-pollution
e incentivos contra la contaminación
حوافز لمكافحة التلوث

209 antisense (strand)
f anti-sens
e sentido inverso
مقاوم للحساسية

210 apiculture
f apiculture
e apicultura
تربية النحل

211 applied cloud physics
f physique appliquée des nuages
e física aplicada de las nubes
الفيزياء التطبيقية للسحب

212 appropriate technology
f technologies appropriées
e tecnología apropiada
تكنولوجيا ملائمة

213 approved facility
f installation agréée
e planta autorizada
مرفق معتمد

214 aquaculture
f aquiculture
e acuicultura
تربية المائيات

215 aquatic acidification
f acidification des eaux
e acidificación de las aguas
تحمض الماء

216 aquatic biota
f biotes aquatiques
e biotas acuáticas
النباتات والحيوانات المائية

217 aquatic food web
f réseau alimentaire aquatique
e red alimentaria acuática
شبكة غذائية مائية

218 aquatic mammals
f mammifères aquatiques
e mamíferos acuáticos
ثدييات مائية

219 aquatic microorganisms
f micro-organismes aquatiques
e micoorganismos acuáticos
كائنات حية دقيقة مائية

220 aquatic plants
f plantes aquatiques
e plantas acuáticas
نباتات مائية

221 aquatic recreational amenities
f aménagements aquatiques de plaisance
e instalaciones acuáticas de recreo
منافع الترويح المائية

222 aqueous cleaner
f produit de nettoyage à l' eau
e limpiador acuoso
منظف بالماء

223 aquifer recharge area
f aire d' alimentation des nappes phréatiques
e zona de alimentación del acuífero
منطقة تغذية المياه الجوفية

224 arable land
f terres arables
e tierras laborables
ارض زراعية

225 arboretum
f arboretum
e arboreto
مستنبت حديقة أشجار

226 archipelagoes
f archipels
e archipiélagos
أرخبيلات

227 architecture
f architecture
e arquitectura
هندسة المبانى

228 arctic ecosystems
f écosystèmes arctiques
e ecosistemas árticos
النظم الايكولوجية للقطب الشمالي

229 arctic region
f régions arctiques
e región ártica
منطقة القطب الشمالي

230 area of environmental stress
f région agressée par la pollution
e zona de tensión ambiental
منطقة اجهاد بيئي

231 arid land ecosystems
f écosystèmes de terres arides
e ecosistemas de tierras áridas
النظم الإيكولوجية للأراضى القاحلة

232 arid lands
f terres arides
e zonas áridas
أراضي قاحلة

233 aromatic
f aromatique
e aromático
عطري

234 arthropods
f arthropodes
e artrópodos
مفصليات

235 artificial drainage
f drainage artificiel
e desagüe artificial
صرف اصطناعي

236 artificial rainfall
f pluie artificielle
e lluvia artificial
مطر اصطناعي

237 asbestos
f amiante
e asbestos
أسبستوس

238 ash-tree
f frêne
e fresno
شجرة الدردار

239 at risk
f à risque
e expuesto a riesgos
معرض للخطر

240 atmosphere
f atmosphère
e atmósfera
الجو

241 atmospheric absorption
f absorption atmosphérique
e absorción atmosférica
امتصاص جوي

242 atmospheric acidity
f acidité de l' air
e acidez de la atmósfera
حموضة جوية

243 atmospheric carbon
f carbone atmosphérique
e carbono atmosférico
الكربون الجوي

244 atmospheric chemistry
f chimie de l' atmosphère
e química de la atmósfera
كمياء الغلاف الجوي

245 atmospheric circulation
f circulation atmosphérique
e circulación atmosférica
دورة جوية

246 atmospheric climate
f climat atmosphérique
e clima atmosférico
المناخ الجوي

247 atmospheric components
f composants atmosphériques
e componentes atmosféricos
مكونات جوية

248 atmospheric composition
f composition atmosphérique
e composición atmosférica
تكوين جوى

249 atmospheric deposition
f dépôt atmosphérique
e deposición atmosférica
ترسب جوي

250 atmospheric disturbance
f perturbation atmosphérique
e perturbación atmosférica
اضطراب جوي

251 atmospheric dynamics
f dynamique de l'atmosphère
e dinámica de la atmósfera
ديناميات الغلاف الجوي

252 atmospheric environment
f milieu atmosphérique
e medio atmosférico
بيئة جوية

253 atmospheric fallout
f retombées atmosphériques
e precipitación atmosférica
تساقط جوي

254 atmospheric models
f modèles atmosphériques
e modelos atmosféricos
نماذج جوية

255 atmospheric monitoring
f surveillance de l'atmosphère
e vigilancia atmosférica
رصد جوى

256 atmospheric ozone
f ozone de l' atmosphère
e ozono atmosférico
كمية الأوزون الجوي

257 atmospheric ozone column
f colonne atmosphérique d' ozone
e columna atmosférica de ozono
عمود الأوزون في الغلاف الجوي

258 atmospheric particulates
f particules atmosphériques
e partículas atmosféricas
جسيمات جوية

259 atmospheric physics
f physique atmosphérique
e física atmosférica
فيزياء جوية

260 atmospheric pollution network
f réseau de surveillance de la pollution atmosphérique
e red de vigilancia de la contaminación atmosférica
شبكة رصد تلوث الغلاف الجوي

261 atmospheric precipitation
f précipitation atmosphérique
e precipitación atmosférica
تساقط جوي

262 atmospheric processes
f phénomènes atmosphériques
e procesos atmosféricos
عمليات جوية

263 atmospheric sciences
f sciences de l' atmosphère
e ciencias de la atmósfera
علوم الغلاف الجوي

264 atmospheric sounder
f sonde atmosphérique
e sonda atmosférica
مسبار جوي

265 atmospheric transfer
f transfert atmosphérique
e transferencia atmosférica
انتقال جوي

266 atmospheric turbidity
f turbidité atmosphérique
e turbidez atmosférica
تكدر الغلاف الجوي

267 austral spring
f printemps austral
e primavera austral
الربيع الجنوبي

268 austral winter
f hiver austral
e invierno austral
الشتاء الجنوبي

269 autoecological data
f données autoécologiques
e datos autoecológicos
بيانات البيئة الذاتية

270 automation detection
f détection automatique
e detección automática
استكشاف أوتوماتي

271 automobile parking
f stationnement des automobiles
e aparcamiento
مكان وقوف السيارات

272 autotroph
f autotrophe
e autótrofo
ذاتي التغذية

273 autotrophy
f autotrophie
e autotrofia
التغذية الذاتية

274 avalanches
f avalanches
e avalanchas y deslizamientos de tierra
انهيارات ثلجية

275 awareness-building
f sensibilisation
e sensibilización
بناء الوعي

276 azeotrope
f azéotrope
e azeotropo
محلول ثابت الغليان

277 azonal soil
f sol azonal
e suelo azonal
تربة غير ناضجة

B

278 back-crossing
f rétrocroisement
e retrocruce
تهجين رجعي

279 back-scattering
f rétrodiffusion
e retrodispersión
الاستطارة

280 back-up
f soutien
e apoyo
احتياطي

281 backcasting
f analyse rétrospective
e análisis retrospectivo
تحليل رجعي

282 background concentration
f concentration de fond
e concentración de fondo
تركيز طبيعي

283 background measurement
f mesure de base
e medición de base
قياس الأساس

284 background ozone
f ozone de fond
e ozono de fondo
الأوزون الطبيعي

285 background pollution
f pollution de fond
e contaminación de fondo
تلوث طبيعي

286 backscattered radiation
f rayonnement rétrodiffusé
e radiación retrodispersada
أشعة مرتدة بالاستطارة

287 bacteria
f bactéries
e bacteria
بكتيريا

288 bacteriology
f bactériologie
e bacteriología
علم الجراثيم

289 baghouse
f installation de filtres à sac
e cámara de filtros de bolsa
حجرة المرشحات الكيسية

290 balance line system
f dispositif à canalisation d'équilibrage
e sistema de canalización equilibrada
نظام خط التوازن

291 balance of nature
f équilibre écologique
e equilibrio de la naturaleza
توازن الطبيعة

292 balks
f terres non labourées
e tierras incultas
أراض غير محروثة

293 ban on imports of timber
f interdiction frappant les importations de bois
e prohibición de las importaciones de madera
حظر استيراد الأخشاب

294 bark beetle
f scolyte
e gorgojo descortezador
خنافس اللحاء

295 basal layer of epidermis
f couche basale de l'épiderme
e capa basal de la epidermis
الطبقة الأساسية للبشرة

296 base metal
f métal commun
e metal común
فلز بخس

297 baseline case
f situation de référence
e caso de referencia
حالة مرجعية

298 baseline data
f données de base
e datos de referencia
بيانات مرجعية

299 baseline monitoring
f station de référence de surveillance
e vigilancia básica
رصد الخط القاعدى

300 basic food requirements
f besoins alimentaires de base
e necesidades alimenticias básicas
المتطلبات الأساسية من الأغذية

301 basic metal
f métal basique
e metal básico
فلز قاعدي

302 basic reaction
f réaction alcaline
e reacción alcalina
تفاعل قلوي

303 basic solution
f solution basique
e solución alcalina
محلول قاعدي

304 battery disposal
f ramassage des piles
e recogida de pilas eléctricas
تخلص من البطاريات

305 beach erosion
f érosion des plages
e erosión de las playas
تآكل الشاطئ

306 beach nourishment
f entretien de la plage
e sustento de la playa
صيانة الشاطئ

307 bearing capacity
f capacité portante [du sol]
e capacidad de carga
قدرة (التربة) على التحمل

308 bedrock
f roche-mère
e lecho de rocas
صخر القاعدة

309 belt skimmer
f récupérateur à bandes
e recuperador de cinta
مكشطة حزامية

310 benchmark survey
f enquête de référence
e estudio de referencia
استقصاء مرجعي

311 beneficiation
f enrichissement
e beneficio
تجديد الخامات

312 benthic
f benthique
e bentónico
قاعي

313 benthic ecosystems
f écosystèmes benthiques
e ecosistemas bentónicos
النظم الإيكولوجية للأعماق

314 benthos
f benthos
e bentos
الأحياء القاعية

315 bequest value
f valeur de transmission
e valor de legado
القيمة التراثية

316 best available control technology
f meilleures techniques de lutte disponibles
e la mejor tecnología disponible de control
أفضل تكنولوجيا المكافحة المتاحة

317 best environmental practices
f meilleures pratiques environnementales
e las mejores prácticas ecológicas
افضل الممارسات البيئية

318 best practicable means
f meilleurs moyens utilisables
e los mejores medios disponibles
افضل الوسائل العملية

319 best practicable practices
f expérience la plus concluante
e las mejores prácticas posibles
افضل الممارسات العملية

320 best practicable technology
f meilleures techniques utilisables
e la mejor tecnología posible
افضل التكنولوجيات العملية

321 beverage can
f boîte-boisson
e lata de bebida
علبة مشروب

322 **beverage industry**
f industrie des boissons
e industria de bebidas
صناعة المشروبات

323 **bilateral conventions**
f conventions bilatérales
e convenios bilaterales
اتفاقيات ثنائية

324 **bilge waters**
f eaux de cale
e aguas de sentina
مياه آسنة

325 **biochip**
f biopuce
e biochip
رقاقة حيوية

326 **bioconversion**
f bioconversion
e conversión biológica
تحويل احيائي

327 **biodata**
f données biotiques
e datos biológicos
بيانات احيائية

328 **bioelectronics**
f bioélectronique
e bioelectrónica
علم الإلكترونيات البيولوجية

329 **bioengineer**
f bioingénieur
e ingeniero biológo
مهندس بيولوجي

330 **bioethics**
f bioéthique
e bioética
قواعد السلوك في العلوم الأحيائية

331 **biofertilization**
f fécondation par biotechnologie
e biofecundación
تسميد أحيائي

332 **bioindustry**
f bio-industrie
e industria biológica
صناعة التكنولوجيا الأحيائية

333 **bioinformatics**
f bioinformatique
e bioinformática
المعلوماتية الأحيائية

334 **bioproductivity**
f bioproductivité
e productividad biológica
الانتاجية الاحيائية

335 **bioregion**
f biorégion
e región biológica
منطقة احيائية

336 **biosensor**
f biodétecteur
e biodetector
جهاز استشعار احيائي

337 **bioaccumulation**
f bioaccumulation
e bioacumulación
تراكم بيولوجي

338 bioassay
f dosage biologique
e ensayo biológico
اختبار بيولوجي

339 biocenosis
f biocénose
e biocenosis
مجموعة أحيائية

340 biochemical action
f action biochimique
e acción bioquímica
فعل كيميائى أحيائي

341 biochemical processes
f processus biochimiques
e procesos bioquímicos
عمليات كيميائية حيوية

342 biochemistry
f biochimie
e bioquímica
كيمياء حيوية

343 biocide
f biocide
e biocida
مبيد آفات احيائي

344 bioclimatic changes
f changements bioclimatiques
e cambios bioclimáticos
تغيرات بيولوجية مناخية

345 bioclimatology
f bioclimatologie
e bioclimatología
علم المناخ الاحيائي

346 biodegradation
f biodégradation
e biodegradación
تحلل حيوى

347 biodegradation rate
f vitesse de biodégradation
e velocidad de biodegradación
معدل التحلل الاحيائي

348 biodeterioration
f biodétérioration
e deterioro biológico
تدهور أحيائي

349 biodiversity
f diversité biologique
e diversidad biológica
تنوع بيولوجي

350 bioenergetics
f bioénergétique
e bioenergética
علم الطاقة الأحيائية

351 biofilter
f biofiltre
e filtro biológico
مصفاة أحيائية

352 biofuel
f biocarburant
e biocarburante
وقود أحيائي

353 biogas
f biogaz
e biogás
غاز أحيائي

354 biogas plant
f installation de biogaz
e fábrica de biogás
مصنع غاز أحيائي

355 biogeocenosis
f biogéocénose
e biogeocenosis
مجموعة أرضية أحيائية

356 biogeochemical cycle
f cycle biogéochimique
e ciclo biogeoquímico
دورة كيميائية أرضية أحيائية

357 biogeochemistry
f biogéochimie
e biogeoquímica
كيمياء أرضية حيوية

358 biogeography
f biogéographie
e biogeografía
جغرافيا إحيائية

359 biological action
f action biologique
e acción biológica
فعل أحيائي

360 biological amplification factor
f coefficient biologique d'incidence
e factor de amplificación biológica
عامل التضخيم الأحيائي

361 biological balance
f équilibre biologique
e equilibrio biológico
توازن أحيائي

362 biological control
f lutte biologique
e lucha biológica
مكافحة أحيائية

363 biological control agent
f agent de lutte biologique
e agente de lucha biológica
عامل مكافحة أحيائية

364 biological control of pests
f lutte biologique contre les parasites
e control biológico de plagas
مكافحة بيولوجية للآفات

365 biological cycle
f cycle biologique
e ciclo biológico
دورة أحيائية

366 biological diversity
f biodiversités
e biodiversidad
تنوع بيولوجي

367 biological diversity and protected areas
f biodiversité et zones protégées
e biodiversidad y regiones protegidas
التنوع البيولوجي والمناطق المحمية

368 biological energy conversion
f bioconversion de l' énergie
e conversión biológica de la energía
التحويل الأحيائى للطاقة

369 biological exposure indicator
f indicateur biologique d' exposition
e indicador biológico de la exposición
مؤشر التعرض الأحيائي

370 biological indicator species
f indicateur biologique
e indicador biológico
انواع احيائية مؤشرة

371 biological indicators
f indicateurs biologiques
e indicadores biológicos
مؤشرات بيولوجية

372 biological nitrogen fixation
f fixation biologique de l'azote
e fijación biológica del nitrógeno
التثبيت البيولوجى للنيتروجين

373 biological process
f voie biologique
e proceso biológico
عملية احيائية

374 biological resources
f ressources biologiques
e recursos biológicos
موارد بيولوجية

375 biological strains of organisms
f souches biologiques d'organismes
e cepas biológicas de organismos
السلالات البيولوجية للكائنات العضوية

376 biological treatment
f épuration biologique
e tratamiento biológico
معالجة احيائية

377 biological uptake
f absorption biologique
e absorción biológica
الامتصاص الاحيائي

378 biological weapons
f armes biologiques
e armas biológicas
أسلحة بيولوجية

379 biology
f biologie
e biología
علم الأحياء

380 biomagnification
f bioamplification
e biomagnificación
تضخيم احيائي

381 biomass
f biomasse
e biomasa
كتلة احيائية

382 biomass-based fuels
f combustibles à partir de la biomasse
e combustibles (obtenidos) de biomasa
وقود أحيائى

383 biomass energy
f énergie de la biomasse
e energía de la biomasa
طاقة الكتلة الاحيائية

384 biomass fuel
f biocarburant
e combustible de biomasa
وقود الكتلة الاحيائية

385 biomass hydrocarbon
f biohydrocarbure
e biohidrocarburo
هيدروكربون احيائي

386 biomass plantation
f plantation pour la production de biomasse
e plantación para la producción de biomasa
مزرعة الكتلة الاحيائية

387 biomass supplies
f réserves de biomasse
e reservas de biomasa
موارد الكتلة الاحيائية

388 biome
f biome
e bioma
منطقة احيائية

389 biomonitoring
f surveillance biologique
e vigilancia biológica
رصد احيائي

390 bioresource
f bioressource
e recurso biológico
مورد احيائي

391 biosafety
f prévention des risques biotechnologiques
e seguridad de la biotecnología
السلامة في التكنولوجيا الحيوية

392 biospheric balance
f équilibre de la biosphère
e equilibrio de la biosfera
توازن المحيط الحيوى

393 biota
f biote
e biota
حيويات

394 biotechnological issues
f ressources biotechnologiques
e recursos biotecnológicos
قضايا التكنولوجيا الحيوية

395 biotechnological safety
f prévention des risques biotechnologiques
e seguridad de biotecnología
السلامة في التكنولوجيا الاحيائية

396 biotechnology
f biotechnologie
e biotecnología
تكنولوجيا أحيائية

397 biotic
f biotique
e biótico
أحيائى

398 biotic control
f lutte par des moyens biotiques
e control biótico
مكافحة احيائية

399 biotope
f biotope
e biotopo
موئل أحيائى

400 birds
f oiseaux
e aves
طيور

401 birth control
f régulation de naissances
e control de la natalidad
تحديد النسل

402 bisphere reserve
f réserve de la biosphère
e reserva de biosfera
محتجز الغلاف الحيوي

403 blast furnaces
f hauts fourneaux
e altos hornos
أفران الصهر

404 body of water
f masse d'eau
e masa de agua
جسم مائي

405 body tissue
f tissus biologiques
e tejidos orgánicos
انسجة احيائية

406 bog
f tourbière
e turbera
مستنقع

407 bolt-on technology
f technique facilement adaptable
e tecnología fácil de adaptar
تكنولوجيا سهلة التطبيق

408 border effect
f effet de lisière
e efecto de linde
اثر حدودي

409 boreal country
f pays de la zone boréale
e país del hemisferio boreal
بلد شمالي

410 boreal forest
f forêt boréale
e bosque boreal
غابة شمالية

411 botanical gardens
f jardins botaniques
e jardines botánicos
حدائق نباتية

412 botany
f botanique
e botánica
علم النبات

413 bottom-living fish
f poissons de fond
e pez bentónico
اسماك قاعية

414 bottom-up planning
f planification à partir de la base
e planificación "de abajo arriba"
تخطيط شامل

415 bound nitogen
f azote fixé
e nitrógeno combinado
نتروجين مثبت

416 boundary layer
f couche limite
e capa límite
طبقة فاصلة

417 brackish water
f eau saumâtre
e agua salobre
ماء اجاج

418 branched
f ramifié
e ramificado
متفرع

419 breakdown of wastes
f dégradation des déchets
e degradación de desechos
تحلل النفايات

420 breeding area
f zone de reproduction
e zona de reproducción
منطقة تكاثر

421 breeding ground
f lieu de couvaison
e zona de reproducción
مكان الانسال

422 brewing industry
f brasserie
e industria cervecera
صناعة التخمير

423 bridge technology
f technique de transition
e técnica de transición
تكنولوجيا انتقالية

424 broadleaved forest
f forêt de feuillus
e bosque de frondosas
غابة عريضة الاوراق

425 brominated species
f produit bromé
e producto bromado
منتجات محتوية على البروم

426 bromination
f bromation
e bromación
معالجة بالبروم

427 brown coal
f lignite
e lignito
فحم بني

428 budwood
f greffons
e injerto
اغصان التطعيم

429 buffer solution
f solution tampon
e solución amortiguadora
محلول منظم

430 build-up (of pollutants)
f accumulation (des polluants)
e acumulación (de contaminantes)
تراكم الملوثات

431 building materials
f matériaux de construction
e materiales de construcción
مواد بناء

432 building poles
f bois de construction
e postes de construcción
اعمدة البناء

433 building technology
f technologie du bâtiment
e tecnología de la edificación
تكنولوجيا المبانى

434 built drainage systems
f systèmes bâtis de drainage
e sistemas artificiales de drenaje
شبكات تصريف مبنية

435 built environment
f cadre bâti
e zonas edificadas
بيئة مبنية

436 built structures
f structures construites
e estructuras construidas
إنشاءات مبنية

437 built-in biological barrier
f barrière biologique intégrée
e barrera biológica incorporada
حاجز احيائي داخلي

438 built-up areas
f zones bâties
e zonas construídas
مناطق مبنية

439 bulk deposition
f dépôt en masse
e deposición en masa
ترسب بكميات كبيرة

440 burden sharing
f partage des obligations
e distribución de obligaciones
تقاسم الاعباء

441 burnt lime
f chaux calcinée
e cal calcinada
جير حي

442 bush fire
f feu de broussailles
e incendio de matorrales
حريق الادغال

443 buy-back facility
f centre de rachat
e centro de compra
مرفق اعادة الشراء

444 by-catch
f prise accessoire
e captura incidental
الصيد العرضي

C

445 cadmium contamination
f contamination par le cadmium
e contaminación con cadmio
تلوث بالكادميوم

446 calibrated fluxes
f valeurs étalonnées des flux
e flujos calibrados
التدفقات المعايرة

447 calibration
f étalonnage
e calibración
معايرة

448 calibration of measuring equipment
f calibrage des appareils de mesure
e calibración del equipo de medición
معايرة معدات القياس

449 candidate vaccine
f vaccin au stade expérimental
e vacuna experimental
لقاح تجريبي

450 canopy density
f densité du couvert (forestier)
e densidad de la cubierta de copas
كثافة الظلة

451 canopy manipulation
f aménagement du couvert
e manipulación de la cubierta de copas
التأثير في الظلة

452 cap
f maximum
e máximo
حد اعلى

453 capacity building
f constitution de capacité
e creación de capacidad
بناء القدرات

454 capital accumulation
f accumulation de capital
e acumulación de capital
تجميع رأس المال

455 captive propagation
f reproduction en captivité
e reproducción en cautiverio
تكاثر الحيوانات الحبيسة

456 carbon-fixing tree
f arbre fixant le carbone
e árbol que fija el carbon
شجرة مثبتة للكربون

457 carbon canister
f cartouche de carbone
e cartucho de carbono
وعاء كربون

458 carbon cycle
f cycle du carbone
e ciclo del carbono
دورة الكربون

459 carbon material
f matière carbonée
e materia carbónica
مادة كربونية

460 carbon offsets
f contrepartie de la fixation du carbone
e emisiones de carbono
تعويض الكربون

461 carbon sink
f puits de carbone
e sumidero del carbono
بالوعة كربون

462 carbon soot
f suie
e hollín
سناج الكربون

463 carcinogens
f substances cancérigènes
e cancerígenos
مسببات السرطان

464 cardiology
f cardiologie
e cardiología
علم أمراض القلب

465 cartography
f cartographie
e cartografía
علم رسم الخرائط

466 cascade impactor
f conimètre à impact en cascade
e impactor en cascada
جهاز لتجميع الهباء

467 cash crop
f culture de rapport
e cultivo comercial
محصول نقدي

468 catalysis
f catalyse
e catálisis
حفز

469 catalyst
f catalyseur
e catalizador
حافز

470 catalytic chain
f cycle catalytique
e ciclo catalítico
سلسلة حفز

471 catalytic exhaust system
f système d'échappement catalytique
e sistema de escape con convertidor catalítico
نظام حفزي للعادم

472 catalytic ozone destruction cycle
f cycle de destruction catalytique de l'ozone
e ciclo de destrucción catalítica del ozono
دورة الاتلاف الحفزي للاوزون

473 catastrophic phenomena
f phénomènes catastrophiques
e eventos catastróficos
ظواهر الكوارث

474 catch crop
f culture dérobée
e cosecha suplementaria
محصول اضافي سريع

475 catch technologies
f techniques de capture
e técnicas de pesca
تكنولوجيات الصيد

476 catchment
f captation d'eau
e captación
مستجمع مياه

477 catchment basin
f bassin d'alimentation
e cuenca hidrográfica
حوض تجميع

478 cattle grazing
f pacage
e pastoreo
رعي الماشية

479 cattle manure
f bouse de vache
e estiércol de bovinos
سماد طبيعي

480 causative organism
f agent pathogène
e organismo patógeno
كائن مسبب لمرض

481 ceiling value
f plafond
e valor máximo
القيمة القصوى

482 cell line
f lignée cellulaire
e estirpe celular
سلالة خلوية

483 cell proliferation factor
f facteur de prolifération cellulaire
e factor de proliferación celular
عامل تكاثر الخلايا

484 cellular plastic
f plastique alvéolaire
e plástico celular
لدائن خلوية

485 cement industry
f industrie du ciment
e industria del cemento
صناعة الأسمنت

486 certification requirements
f régles d'homologation
e requisitos de certificación
متطلبات منح الشهادات

487 changes in land use
f évolution de l'utilisation des sols
e cambios en el uso de la tierra
تغيرات في استخدام الاراضي

488 changing atmosphere
f atmosphère en évolution
e cambios en la atmósfera
غلاف جوي متغير

489 **charges on water pollution**
f redevances de pollution des eaux
e recargo por contaminación del agua
رسوم على تلويث المياه

490 **chemi-thermo-mechanical pulping**
f pâte chimiothermo-mécanique
e pasta quimiotermomecánica
استخلاص كيميائي-حراري ميكانيكي للب الورق

491 **chemical barrier**
f barrière chimique
e barrera química
حاجز كيميائي

492 **chemical burn**
f brûlure chimique
e quemadura química
حرق كيميائي

493 **chemical change**
f transformation(s) chimique(s)
e transformación química
تغير كيميائي

494 **chemical cleaning of coal**
f épuration chimique du charbon
e purificación química del carbón
تنقية الفحم كيميائيا

495 **chemical decontamination**
f décontamination chimique
e descontaminación química
إزالة التلوث الكيميائى

496 **chemical dump**
f décharge de déchets chimiques
e vertedero de desechos químicos
نفايات كيميائية

497 **chemical emergency**
f alerte chimique
e situación de emergencia química
حالة طوارئ كيميائية

498 **chemical engineering**
f génie chimique
e ingeniería química
هندسة كيميائية

499 **chemical oceanography**
f océanographie chimique
e oceanografía química
أوقيانوغرافيا كيميائية

500 **chemical oxygen demand**
f demande chimique en oxygène
e demanda química de oxígeno
الحاجة الكيميائية للأكسجين

501 **chemical poisoning**
f intoxication par des substances chimiques
e envenenamiento con sustancias químicas
تسمم كيميائي

502 **chemical reactivity**
f réactivité chimique
e reactividad química
تفاعلية كيميائية

503 chemical safety
f sécurité des substances chimiques
e prevención de los riesgos químicos
سلامة كيميائية

504 chemical substitute
f substance chimique de substitution
e sustituto químico
بديل كيميائي

505 chemical theory
f hypothèse chimique
e teoría química
النظرية الكيميائية

506 chemical transformation
f transformation chimique
e transformación química
تحول كيميائي

507 chemical transmitter
f télémédiateur chimique
e transmisor químico
افراز كيميائي

508 chemical treatment of waste
f traitement chimique des déchets
e tratamiento químico de desechos
معالجة كيميائية للنفايات

509 chemical weapons
f armes chimiques
e armas químicas
أسلحة كيميائية

510 chemically inert
f chimiquement inerte
e químicamente inerte
خامل كيميائيا

511 chemically reactive substance
f substance chimiquement réactive
e sustancia (químicamente) reactiva
مادة متفاعلة كيميائيا

512 chemistry of the atmosphere
f chimie de l'atmosphère
e química de la atmósfera
كيمياء الغلاف الجوي

513 chemo-organotrophic
f chimio-organotrophique
e quimioorganotrófico
تغذية كيميائية – عضوية

514 chemoautotroph
f chimiotrophe
e quimioautótrofo
ذاتي التغذية كيميائيا

515 chimneys
f cheminées
e chimeneas
مداخن

516 chloride prescrubber
f dispositif de prélavage des chlorures
e prelavador de cloruros
جهاز الغسل الاولى بالكلوريد

517 **chlorinated**
f chloré
e clorado
مكلور

518 **chlorination**
f chloration
e cloración
كلورة

519 **chlorine**
f chlore
e cloro
كلور

520 **chlorine sink**
f puits de chlore
e sumidero del cloro
بالوعة كلور

521 **chlorine species**
f composés chlorés
e compuestos clorados
مركبات الكلور

522 **chlorine system**
f cycle du chlore
e ciclo del cloro
دورة الكلور

523 **chlorinolysis**
f chlorolyse
e clorolisis
انحلال مركبات الكلور

524 **chromatographic analysis**
f analyse chromatographique
e análisis cromatográfico
تحليل بالفصل اللونى

525 **civil engineering**
f génie civil
e ingeniería civil
هندسة مدنية

526 **clay adsorption**
f adsorption sur argile
e adsorción en arcilla
امتزاز الطين

527 **clean air**
f air pur
e aire puro
هواء نقي

528 **clean coal technology**
f techniques de charbon épuré
e técnicas poco o menos contaminantes de uso del carbón
تكنولوجيا الفحم النظيفة

529 **clean industries**
f industries propres
e industrias poco o menos contaminantes
صناعات نظيفة

530 **clean product**
f produit propre
e producto no contaminante
منتج نظيف

531 **clean rain**
f pluie propre
e lluvia limpia
مطر نظيف

532 clean technologies
f techniques non polluantes
e tecnologías o técnicas poco o menos contaminantes
تكنولوجيات نظيفة

533 clean water
f eau salubre
e agua limpia
مياه نظيفة

534 cleaner technologies
f techniques propres
e tecnologías poco o menos contaminantes
تكنولوجيات انظف

535 cleaning of the air
f épuration de l'air
e purificación del aire
تنقية الهواء

536 cleaning solvent
f solvant de dégraissage
e solvente de limpieza
محلول تنظيف

537 clearing house
f centre d'échange
e mecanismo de selección y financiación de proyectos
غرفة مقاصة

538 climate alert
f alerte climatique
e alerta climático
انذارمناخي

539 climate applications
f applications de climatologie
e aplicaciones de la climatología
تطبيقات مناخية

540 climate change
f changement(s) climatique(s)
e cambios del clima
تغير المناخ

541 climate cycle
f cycle climatique
e ciclo del clima
دورة المناخ

542 climate diagnostic
f analyse du climat
e conclusión sobre el estado del clima
تحليل المناخ

543 climate indicator
f témoin des climats
e indicador (del estado) del clima
دليل مناخي

544 climate model
f modèle climatique
e modelo del clima
نموذج مناخي

545 climate monitoring
f surveillance du climat
e vigilancia del clima
رصد المناخ

546 climate observing station
f station d'observation climatique
e estación de observación del clima
محطة مراقبة المناخ

547 climate protection
f protection du climat
e protección del clima
حماية المناخ

548 climate response
f réaction du climat
e reacción del clima
استجابة المناخ

549 climate science
f climatologie
e climatología
علم المناخ

550 climate sensitivity
f sensibilité du climat
e sensibilidad del clima
حساسية مناخ

551 climate system
f système climatique
e sistema climático
نظام مناخي

552 climate warming
f réchauffement du climat
e calentamiento del clima
احترار المناخ

553 climate watch
f veille climatologique
e vigilancia del clima
تتبع المناخ

554 climate-induced changes
f évolution sous l'effet climat
e cambios de origen climático
تغيرات متأثرة بالمناخ

555 climatic anomaly
f anomalie climatique
e anomalía climática
شذوذ مناخي

556 climatic atlas
f atlas climatique
e atlas del clima
اطلس مناخي

557 climatic cycles
f cycles climatiques
e ciclos climáticos
دورات مناخية

558 climatic disaster
f catastrophe climatique
e desastre climático
كارثة مناخية

559 climatic divide
f frontière climatique
e frontera climática
فاصل مناخي

560 climatic element
f élément climatique
e elemento climático
عنصر مناخي

561 climatic event
f phénomène climatique
e fenómeno climático
ظاهرة مناخية

562 climatic factors
f facteurs climatiques
e factores climáticos
عوامل مناخية

563 climatic forecast
f prévision climatique
e pronóstico del estado del clima
تنبؤ مناخي

564 climatic hazard
f risque lié au climat
e riesgo de origen climático
خطر مناخي

565 climatic model
f modèle climatique
e modelo del clima
نموذج مناخي

566 climatic record
f relevé climatologique
e dato climatológico
سجل مناخي

567 climatic region
f région climatique
e región climática
اقليم مناخي

568 climatic shift
f modification climatique
e variación del clima
تحول مناخي

569 climatic stress
f contraintes climatiques
e tensiones debidas al clima
اجهاد مناخي

570 climatic upheaval
f bouleversement climatique
e trastorno climático
اضطراب مناخي

571 climatic variability
f variabilité du climat
e variabilidad del clima
التقلبية المناخية

572 climatic zone
f zone climatique
e zona climática
منطقة مناخية

573 climatography
f climatographie
e climatografía
علم رسم المناخ

574 climatological forecast
f prévision climatologique
e predicción climatológica
تنبؤ مناخي

575 climatological station
f station climatologique
e estación climatólogica
محطة مناخية

576 climatologist
f climatologiste
e climatólogo
عالم مناخ

577 climatology
f climatologie
e climatología
علم المناخ

578 climax
f climax
e clímax
ذروة

579 climax forest
f forêt climatique
e bosque en equilibrio ecológico
غابة قمة

580 cloning
f clonage
e clonación
استنساخ

581 closed-cup test
f essai en creuset fermé
e prueba en crisol cerrado
اختبار الكأس المغلق

582 closed forest
f forêt dense
e bosque denso
غابة ممتلئة

583 closed forest cover
f couvert forestier dense
e cubierta densa de copas
غطاء حرجي ممتلئ

584 closed wood
f bois en défens
e veda de pastoreo
غابة مغلقة

585 closeness
f densité
e densidad de masa
الكثافة

586 cloud albedo
f albédo des nuages
e albedo de las nubes
بياض السحب

587 cloud amount
f nébulosité (degrè de)
e grado de nebulosidad
نسبة التغيم

588 cloud climatology
f climatologie des nuages
e climatología de las nubes
علم مناخ السحب

589 cloud droplets
f gouttelettes de nuages
e gotas de las nubes
قطيرات السحب

590 cloud field
f champ de nuages
e campo de nubes
مجال السحب

591 cloud physics
f physique des nuages
e física de las nubes
فيزياء السحب

592 cloud process
f processus de nébulosité
e proceso de formación de las nubes
عملية تكون السحب

593 cloud-top height
f hauteur du sommet des nuages
e altura de la cima de la nube
ارتفاع قمة السحب

594 cloudiness
f nébulosité
e nebulosidad
التغيم

595 cluster
f rubrique
e grupo
مجموعة

596 coal beneficiation
f valorisation du charbon
e valorización del carbón
اغناء الفحم

597 coal cleaning plan
f station d'épuration du charbon
e instalación de depuración del carbón
مصنع تنظيف الفحم

598 coal equivalent
f équivalent charbon
e equivalente en carbón
مكافئ فحمي

599 coal gasification
f gazéification du charbon
e gasificación del carbón
تحويل الفحم إلى غاز

600 coal liquefaction
f liquéfaction du charbon
e licuefacción del carbón
إسالة الفحم

601 coastal waters
f eaux côtières
e aguas costeras
مياه ساحلية

602 coastal zone management
f gestion des zones côtières
e ordenación de las zonas costeras
ادارة المناطق الساحلية

603 coastal area
f zone littorale
e zona costera
منطقة ساحلية

604 coastal biodiversity
f diversité biologique des zones côtières
e diversidad biológica de las zonas costeras
التنوع البيولوجي الساحلي

605 coastal climate
f climat littoral
e clima costero
مناخ ساحلي

606 coastal commons
f patrimoine côtier
e patrimonio costero
المشاعات الساحلية

607 coastal development
f aménagement du littoral
e desarrollo de las zonas costeras
تنمية السواحل

608 coastal evolution
f évolution des zones littorales
e evolución de las costas
تطور السواحل

609 coastal fog
f brouillard côtier
e niebla costera
ضباب ساحلي

610 coastal margins
f bande côtière
e franja costera
حواف ساحلية

611 **coastal protected area**
f zone littorale protégée
e zona costera protegida
منطقة ساحلية محمية

612 **coastal shelf**
f plateau continental
e plataforma continental
جرف قاري

613 **coastal water**
f eaux côtières
e aguas costeras
مياه ساحلية

614 **coat protein**
f protéine d'enveloppe
e proteína de cubierta
بروتين مغلف

615 **coating**
f produit de surfaçage
e revestimiento
طلاء

616 **coefficient of haze**
f coefficient de brume
e coeficiente de neblina
معامل الرهج

617 **cogeneration**
f cogénération
e cogeneración
توليد مشترك

618 **coliform count**
f numération des coliformes
e contenido de coliformes
عدد البكتريا المعدية

619 **collecting system**
f adduction
e sistema colector
نظام تجميع

620 **collection of household refuse**
f enlèvement des ordures ménagères
e recolección de desperdicios domésticos
جمع القمامة المنزلية

621 **collective use rights**
f droits d'usage collectif
e derechos de uso colectivo
حقوق الاستخدام الجماعي

622 **colonizing species**
f espèce colonisatrice
e especies colonizadoras
انواع مستوطنة

623 **column of ozone**
f colonne atmosphérique d'ozone
e columna (atmosférica) de ozono
عمود اوزون

624 **combating climate change**
f lutte contre les changements climatiques
e lucha contra los cambios climáticos
مكافحة تغير المناخ

625 **combustion emissions**
f émissions dues à la combustion
e emisiones poducidas por la combustión
انبعاثات الاحتراق

626 combustion plant
f installation de combustion
e instalación de combustión
منشأة حرق

627 combustion source
f foyer
e fuente de combustión
مصدر احتراق

628 combustor
f chambre de combustion
e cámara de combustión
غرفة احتراق

629 commercial applications
f applications commerciales
e aplicaciones comerciales
تطبيقات تجارية

630 commercial noise
f bruits commerciaux
e ruido comercial
ضوضاء تجارية

631 commercially available substitute
f substitut disponible sur le marché
e producto sustitutivo disponible en el mercado
بديل متوفر تجاريا

632 common name
f nom usuel
e nombre común
اسم عام

633 community architecture
f architecture de quartier
e arquitectura comunitaria
الهندسة المجتمعية

634 community rights
f droits communautaires
e derechos comunitarios
حقوق المجتمع

635 compacted soil
f sol battu
e suelo compactado
تربة متراصة

636 compaction
f compactage
e compactación
رص

637 competitive ability
f capacité compétitive
e capacidad competitiva
قدرة تنافسية

638 compliance
f conformité
e cumplimiento
امتثال

639 component
f constituant
e componente
عنصر ؛ مكون

640 composite pollution
f pollution composée
e contaminación compuesta
تلوث مركب

641 compound organic matter
f matière organique complexe
e materia orgánica compuesta
مادة عضوية مركبة

642 compressed natural gas
f gaz naturel comprimé
e gas natural comprimido
غاز طبيعي مضغوط

643 compression ratio
f taux de compression
e grado de compresión
نسبة الضغط

644 compressive resistance
f résistance à la compression
e resistencia a la compresión
مقاومة الضغط

645 compulsory measures
f mesures contraignantes
e medidas obligatorias
تدابير الزامية

646 compulsory return
f retour obligatoire
e devolución obligatoria
اعادة الزامية

647 concentration basin
f bassin de concentration
e cuenca de concentración
حوض التركيز

648 concentration factor
f facteur de concentration
e factor de concentración
عامل التركيز

649 concentration level
f concentration
e grado de concentración
مستوي التركيز

650 concessional terms
f conditions favorables
e condiciones de favor
شروط تساهلية

651 concessionality
f caractère favorable
e carácter concesionario
التساهلية

652 condensate
f condensat
e condensado
مادة مكثفة

653 coniferous forest
f forêt de résineux
e bosque de coníferas
غابة صنوبرية

654 coniferous saw-wood
f sciages résineux
e madera de coníferas
خشب الصنوبر المنشور

655 coning
f panache conique
e penacho cónico
اتخاذ شكل مخروطي

656 conservation of biological diversity
f préservation de la diversité biologique
e conservación de la diversidad biológica
حفظ التنوع البيولوجي

657 conservation of genetical diversity
f préservation de diversité génétique
e conservación de la diversidad genética
حفظ التنوع الجيني

658 conservation of nature
f protection de la nature
e conservación de la naturaleza
حفظ الطبيعة

659 conservation of soil
f protection des sols
e conservación del suelo
حفظ التربة

660 conservationist
f écologiste
e ecologista
داعية لحفظ الصبيعة

661 consignment (of waste)
f expédition (de déchets)
e transporte (de desechos)
شحن النفايات

662 constant pressure chart
f carte à pression constante
e mapa de presión constante
خريطة الضغط الثابت

663 constituencies
f groupes cible
e grupos interesados
فئات معنية

664 consumer non-durable (good)
f bien de consommation courant
e bien de consumo no duradero
بضائع استهلاكية

665 consumptive user
f utilisateur-consommateur
e usuario consumidor
مستعمل مبذر

666 consumptive values
f valeurs de consommation
e valores de consumo
قيم استهلاكية

667 contained facility
f installation d'utilisation confinée
e instalación de trabajo confinado
مرفق معزول

668 contained use
f usage confiné
e uso confinado
استخدام معزول

669 containment
f confinement
e confinamiento
حصر استخدام

670 containment categories
f catégories de confinement
e categorías de confinamiento
فئات العزل

671 **containment of controlled substances**
f confinement des substances réglementées
e confinamiento de (las) sustancias controladas
حصر استخدام المواد الخاضعة للرقابة

672 **containment of solid wastes**
f confinement des déchets solides
e confinamiento de los desechos sólidos
احتواء النفايات الصلبة

673 **contaminant**
f contaminant
e contaminante
ملوث

674 **content**
f teneur (en)
e contenido
محتوى

675 **continental ice sheet**
f calotte glaciaire continentale
e manto de hielo continental
غطاء جليدي قاري

676 **continuous sampling**
f échantillonnage en continu
e muestreo continuo
اخذ عينات متواصل

677 **control technologies**
f technologies de maîtrise de la pollution
e tecnologías de lucha
تكنولوجيات المكافحة

678 **controlled dumping**
f décharge contrôlée
e descarga controlada
القاء خاضع للرقابة

679 **controlled landfilling**
f mise en décharge contrôlée
e descarga controlada de desechos en vertederos
دفن القمامة الخاضع للرقابة

680 **controlled oxidation**
f oxydation ménagée
e oxidación controlada
اكسدة محكومة

681 **conventional fossil fuels**
f combustibles fossiles traditionnels
e combustibles fósiles tradicionales
انواع الوقود الحفرى التقليدية

682 **conventional petrol engine**
f moteur à essence classique
e motor de gasolina convencional
محرك تقليدي يعمل بالبنزين

683 **conventional resistance**
f résistance non induite
e resistencia convencional
مقاومة تقليدية

684 **conversion time**
f temps de mise au point
e tiempo de conversión
الزمن اللازم للتحويل

685 **cooking fuel**
f combustible pour la cuisson des aliments
e combustible de cocina
وقود الطبخ

686 **coolant**
f liquide réfrigérant
e refrigerante
مبرد

687 **cooling water**
f eau de refroidissement
e agua de refrigeración
ماء التبريد

688 **coppice forest**
f taillis
e tallar
غابة اشجار صغيرة

689 **coral bleaching**
f décoloration des coraux
e descoloramiento de los corales
تبييض المرجان

690 **coral reef**
f récif de corail
e arrecife de coral
شعب مرجانية

691 **cordwood**
f bois de corde
e rollizos (para leña)
حطب

692 **core**
f carotte
e testigo
عينة جوفية اسطوانية

693 **cosmic radiation**
f rayonnement cosmique
e radiación cósmica
اشعاع كوني

694 **cost internalization**
f internalisation des coûts
e internalización de los costos
تدخيل التكلفة

695 **cost recovery scheme**
f dispositif de récupération des coûts
e plan de recuperación de los costos
خطة استرداد التكاليف

696 **cost impact**
f effet(s) de coût
e costos resultantes
اثر التكلفة

697 **cost-benefit analysis**
f analyse coût-avantage
e análisis costo-beneficio
تحليل التكلفة والفائدة

698 **cost-effective**
f efficace en termes de coût
e eficaz en función del costo
فعال الكلفة

699 **cost-effectiveness**
f rapport coût-efficacité
e relación costo-eficacia
فعالية التكلفة

700 cost-effectiveness analysis
f analyse coût-efficacité
e análisis de la relación costo-eficacia
تحليل فعالية التكلفة

701 cost-offsetting advantage
f avantage en contrepartie du coût
e ventaja que compensa el costo
ميزة معادلة التكلفة

702 cost-optimal
f optimal en termes de coût
e óptimo desde el punto de vista del costo
امثل من حيث التكلفة

703 counter-claim
f demande reconventionnelle
e contrademanda
ادعاء مقابل

704 cradle to grave management
f gestion de la totalité du cycle de vie
e gestión de la cuna a la sepultura
ادارة "من المهد الى اللحد"

705 creeping pollution
f pollution rampante
e contaminación lenta
تلويث زاحف

706 crime against the environment
f crime contre l'environnement
e delito contre el medio ambiente
جريمة ضد البيئة

707 critical load
f charge critique
e carga crítica
حمل حرج

708 critical loads approach
f méthode des charges critiques
e método de las cargas críticas
نهج الاحمال الحرجة

709 crop biotechnology
f biotechnologie agricole
e biotecnología agrícola
التكنولوجيا الاحيائية الزراعية

710 crop coefficient
f coefficient culturel
e coeficiente de cultivo
معامل محصولي

711 crop damage
f dommages causés aux cultures
e daños causados a los cultivos
تلف المحصول

712 crop failure
f mauvaise récolte
e mala cosecha
فقد المحصول

713 crop pest
f parasites agricoles
e plagas agrícolas
آفة زراعية

714 crop plant
f plante cultivée
e planta cultivada
نبات محصولي

715 **crop waste**
f déchets de récolte
e desechos de corrales
نفايات المحاصيل

716 **crop weeding**
f désherbage
e escarda
اقتلاع الاعشاب من المحصول

717 **croplands**
f terres cultivables
e tierras de cultivo
اراض زراعية

718 **cropping season**
f récoltes
e estación de la cosecha
موسم الحصاد

719 **cropping system**
f système de culture
e sistema de cultivo
نظام زراعة

720 **cross-border pollution**
f pollution transfrontière
e contaminación transfronteriza
تلوث عابر للحدود

721 **cross-cutting issue**
f problème global
e cuestión intersectorial
قضية شاملة

722 **cross-media pollution**
f pollution touchant plusieurs milieux
e contaminación de varios medios
تلوث شامل لعدة أوساط

723 **cross-section**
f section efficace
e sección eficaz
مقطع عرضي

724 **cyclones**
f cyclones
e ciclones
أعاصير

725 **cyclone separation**
f cyclonage
e separación ciclónica
فرز دوامي

726 **cyclonic energy**
f énergie des cyclones
e energía de los ciclones
طاقة اعصارية

727 **cytogenetics**
f cytogenétique
e citogenética
علم الوراثة الخلوي

728 **cytokinesis**
f cytocinèse
e citocinesis
تغيرات مصاحبة لتطور

729 **cytology**
f cytologie
e citología
علم حياة وتكوين الخلايا

D

730 daily variation
f variation diurne
e variación diurna
تباين يومي

731 damage assessment
f évaluation des dégâts
e evaluación de daños
تقييم الضرر

732 damage to the environment
f atteinte à l'environnement
e daños al medio ambiente
ضرر للبيئة

733 damping
f atténuation
e parada temporal
توهين

734 danger label
f étiquette de danger
e etiqueta de peligro
علامة الخطر

735 danger level
f niveau d'alarme
e nivel de peligro
مستوى الخطر

736 dangerous goods
f marchandises dangereuses
e mercaderías peligrosas
بضائع خطرة

737 data profile
f fiche descriptive
e perfil de datos
موجز بيانات

738 data strategy
f stratégie de collecte d'information
e estrategia de reunión de datos
استراتيجية جمع البيانات

739 de-duster
f dépoussiéreur
e captador de polvo
مزيل الغبار

740 dead lime
f chaux éteinte
e cal apagada
جير مطفأ

741 debris
f détritus
e detritos
ركام

742 debt for development swap
f conversion de créances pour le financement de projets de développement
e canje de deuda por proyectos de desarrollo
تحويل الديون لتمويل التنمية

743 debt for sustainable development swap
f conversion de créances pour le financement d'un développement durable
e canje de deuda por financiación del desarrollo sostenible
تحويل الدين لتمويل التنمية المستدامة

744 decadal time scale
f échelle de la décennie
e escala cronológica del decenio
جدول حسب العقود الزمنية

745 decay of air pollutants
f désintégration des polluants atmosphériques
e descomposición de contaminantes
تحلل ملوثات الجو

746 decay of organic matter
f décomposition des matières organiques
e descomposición de materia orgánica
تحلل المواد العضوية

747 decay of waves
f amortissement de la houle
e amortiguación de las olas
هدوء الأمواج

748 deciduous forest
f forêt caducifolée
e bosque de especies caducifolias
غابة متساقطة الأوراق

749 decomposer
f décomposeur
e descomponedor
مسبب للتحلل

750 decoupling
f dissociation
e desacoplamiento
فصل

751 decreasing taxa
f espèces en régression
e taxones en disminución
انواع متناقصة الأعداد

752 deep ecology
f écologisme radical
e ecología radical
ايكولوجيا متعمقة

753 deep injection (of wastes)
f injection en profondeur (des déchets)
e inyección profunda (de desechos)
حقن عميق (للنفايات)

754 deep layer injection
f injection en couche profonde
e inyección en capas profundas
حقن فى الطبقات العميقة

755 deep-sea circulation
f circulation des eaux profondes
e circulación de aguas profundas
دورة البحار العميقة

756 deep-sea disposal (of wastes)
f immersion (des déchets) en mer à grande profondeur
e eliminación (de desechos) en aguas profundas
تصريف النفايات في اعماق البحار

757 deep-well injection
f injection en profondeur
e inyección profunda
حقن فى آبار عميقة

758 deeper ocean
f eaux abyssales
e aguas abisales
محيط سحيق

759 deficiency payment
f montant compensatoire
e pago compensatorio
دفعة لتغطية عجز

760 defoliation
f défoliation
e defoliación
سقوط الاوراق

761 deforestation
f déboisement
e deforestación
ازالة الغابات

762 deglaciation
f dégel
e deglaciación
انحسار جليدي

763 dehydrochloration
f déchlorhydratation
e deshidrocloración
ازالة كلور الماء

764 dehydrogenation
f déshydrogénation
e deshidrogenación
ازالة الهيدروجين

765 delayed effect
f effet retardé
e efecto retardado
اثر لاحق

766 deleterious effect
f effet nocif
e efecto nocivo
اثر ضار بالصحة

767 deleterious gas
f gaz délétère
e gas nocivo
غاز ضار

768 deliberate release into the environment
f dissémination volontaire dans l'environnement
e liberación deliberada en el medio ambiente
اطلاق متعمد في البيئة

769 delivery techniques
f techniques d'utilisation
e términos de utilización
تقنيات الاستخدام

770 dendroclimatology
f dendroclimatologie
e dendroclimatología
علم المناخ الشجري

771 denitration
f dénitratation
e desnitración
نزع النيتروجين

772 denitrification
f dénitrification
e desnitrificación
نزع النيترات

773 denitrifying agent
f agent dénitrifiant
e agente desnitrificante
عامل ازالة النيتروجين

774 denitrogenation
f dénitration
e desnitrogenación
ازالة النيتروجين

775 denoxing
f élimination d'oxydes d'azote
e eliminación de óxidos de nitrógeno
نزع اكسيد النيتروجين

776 density
f densité
e densidad
كثافة

777 density currents
f courants de densité
e corrientes de densidad
تيارات الكثافة

778 density of canopy
f densité du couvert forestier
e densidad de la cubierta de copas
كثافة الظلة

779 depletable resource
f ressource non renouvelable
e recurso no renovable
مورد غير متجدد

780 depletion in numbers
f déclin numérique
e decrecimiento numérico
تناقص عددي

781 depletion of soil
f épuisement du sol
e agotamiento del suelo
فقر التربة

782 depletion of the ozone layer
f appauvrissement de la couche d'ozone
e agotamiento de la capa de ozono
استنفاد طبقة الاوزون

783 depletion rate of natural resource stocks
f taux de baisse de stocks de ressources naturelles
e tasa de agotamiento de las reservas de recursos naturales
معدل استنفاد أرصدة الموارد

784 depollution
f dépollution
e descontaminación
ازالة التلوث

785 depolymerization
f dépolymérisation
e depolimerización
ازالة البلمرة

786 deposit refund system
f système de consignation
e sistema de pago y reembolso de depósitos
نظام العربون

787 deposition
f dépôt(s)
e deposición
ترسب

788 deposition velocity
f vitesse de dépôt
e velocidad de deposición
سرعة الترسب

789 deprivation
f dénuement
e carencia
حرمان

790 depthsonde
f sonde de mesure
e sonda de profundidad
سبر الاعماق

791 dermal toxicity
f toxicité par absorption cutanée
e toxicidad dérmica
سمية جلدية

792 descendant of a genetically modified organism
f descendant d'un organisme génétiquement modifié
e descendiente de un organismo genéticamente modificado
سليل كائن محور جينيا

793 desertification control
f lutte contre la désertification
e lucha contra la desertificación
مكافحة التصحر

794 desiccation control
f lutte contre l'aridité
e lucha contra la desecación
مكافحة الجفاف

795 design temperature
f température de calcul
e temperatura prevista en el diseño
درجة الحرارة الاسمية

796 desirable clone
f clone prometteur
e clon deseable
مستنسخ مرغوب فيه

797 desludge
f désenvaser
e desenlodar
ازال الحمأة

798 desoxyribonucleic
f acide désoxyribonucléique (ADN)
e ácido desoxirribonucleico (ADN)
الحمض الخلوي الصبغي

799 destructive exploitation (of species)
f utilisation abusive (des espèces)
e explotación destructiva (de especies)
استغلال متلف (للانواع)

800 desulphurization
f désulfuration
e desulfuración
نزع الكبريت

801 detergency
f détergence
e detergencia
القدرة التنظيفية

802 determination (of a substance)
f dosage (d'une substance)
e determinación
تحديد مادة

803 detoxification
f détoxication
e destoxificación
ازالة السمية

804 devastation of forests
f destruction des forêts
e destrucción forestal
دمار الغابات

805 deviating force
f force déviante
e fuerza de desviación
قوة انحراف

806 dew point
f point de rosée
e punto de rocío
نقطة الندى

807 dewatering
f déshydratation
e deshidratación
نزح الماء

808 diagnostic product
f produit permettant d'établir des diagnostics
e producto para hacer diagnósticos
منتج تشيخصي

809 dielectric (substance)
f diélectrique
e dieléctrico (material)
عازل

810 dielectric oil
f huile isolante
e aceite aislante
زيت عازل

811 differential heating
f échauffement différentiel
e caldeo selectivo
تسخين تبايني

812 diffuse source
f source diffuse
e fuente difusa
مصدر الانتشار

813 diffusion in the air
f diffusion dans l'atmosphère
e difusión en la atmósfera
الانتشار في الهواء

814 digestion tank
f digesteur
e tanque digestor
صهريج هضم

815 digital resolution
f résolution numérique
e resolución digital
تحليل رقمي

816 dimensional stability
f stabilité dimensionnelle
e estabilidad dimensional
ثبات الابعاد

817 dipterocarp forest
f forêt contaminée par diptérocarpacées
e bosque contaminado por dipterocarpáceas
غابة اشجار ثنائية الاوراق

818 direct-use value
f valeur directe d'usage
e valor de uso directo
قيمة الاستخدام المباشر

819 disamenity
f nuisance
e molestia
ضرر

820 disamenity costs
f coûts de la pollution
e costo de las molestias
تكلفة الضرر

821 disaster
f catastrophe
e desastre
كارثة

822 disaster alert
f alerte à la catastrophe
e alerta de desastre
الانذار بالكوارث

823 disaster hazard
f risque de catastrophe
e peligro de desastre
خطر الكارثة

824 disaster preparedness
f préparation aux catastrophes
e preparación para casos de desastre
التأهب للكوارث

825 disaster response
f action en cas de catastrophe
e medidas en caso de desastre
استجابة لكوارث

826 disaster-prone area
f zone à risques
e zona expuesta a desastres
منطقة معرضة للكوارث

827 disaster-prone country
f pays sujet à des catastrophes
e país expuesto a desastres
بلد معرض للكوراث

828 disaster-stricken area
f région sinistrée
e zona afectada por un desastre
منطقة مصابة بالكوارث

829 disc skimmer
f récupérateur a disques
e recuperador con mecanismo de discos
مكشطة قرصية

830 discharge
f rejet
e descarga
تصريف

831 discharge into the environment
f rejet dans l'environnement
e descarga en el medio ambiente
تصريف في البيئة

832 discharge into the sea
f rejet en mer
e descarga en mar
تصريف في البحر

833 discharge monitoring
f surveillance des rejets
e vigilancia de descargas
رصد التصريف

834 discharge permit
f autorisation de rejet
e permiso de descarga
تصريح تصريف

835 discharge pipe
f conduite d'évacuation
e tubo de descarga
أنبوبة تصريف

836 discharge point
f lieu d'émergence
e punto de emergencia
نقطة تصريف

837 discharge rate
f débit d'émission
e tasa de descarga
معدل التصريف

838 discharge standard
f norme de rejet
e norma de descargas
معيار التصريف

839 discharging
f rejet
e acción de descargar
تصريف

840 discolouration (of trees)
f décoloration (des arbres)
e descoloración (de los árboles)
ازالة لون (الاشجار)

841 discomfort
f gêne
e incomodidad
ازعاج

842 disease-free zone
f zone non contaminée
e zona libre de enfermedades
منطقة خالية من الامراض

843 disincentive
f mesure dissuasive
e desincentivo
مثبط

844 disordered environment
f environnement perturbé
e medio ambiente perturbado
بيئة مضطربة

845 dispersal
f libération
e dispersión
تشتيت

846 dispersion in the air
f dispersion dans l'atmosphère
e dispersión en la atmósfera
انتشار في الهواء

847 dispersion of particles
f migration des particules
e dispersión de partículas
تشتت الجزيئات

848 disposable packaging
f emballage perdu
e embalaje desechable
تغليف مستهلك

849 disposable syringe
f seringue jetable
e jeringa desechable
محقن مستهلك

850 disposal at sea
f élimination en mer
e eliminación en el mar
التخلص في البحر

851 disposal contractor
f éliminateur
e contratista de eliminación de desechos
متعهد تصريف

852 disposal in land
f élimination dans le sol
e eliminación en el suelo
الدفن في الارض

853 disposal of wastes
f élimination des déchets
e eliminación de desechos
التخلص من النفايات

854 disposal site
f site d'elimination; décharge
e lugar de eliminación
موقع تصريف

855 disruptive
f qui entraîne des perturbations
e perturbador
معطل

856 dissipation of heat
f dissipation de la chaleur
e disipación del calor
تبديد الحرارة

857 dissociation constant
f constante de dissociation
e constante de disociación
ثابت التفكك

858 dissolved air flotation
f flottation à l'air dissous
e flotación por aire disuelto
تعويم بالهواء المذاب

859 dissolved organic matter
f matière organique dissoute
e materia orgánica disuelta
مادة عضوية مذابة

860 dissolved oxygen
f oxygène dissous
e oxígeno disuelto
اكسجين مذاب

861 distressed area
f zone sinistrée
e zona afectada damnificada
منطقة منكوبة

862 district heating
f chauffage urbain
e calefacción centralizada de barrios
تدفئة المدن

863 disturbed ecological balance
f équilibre écologique perturbé
e equilibrio ecológico alterado
توازن ايكولوجي مضطرب

864 disturbed forest
f forêt en mauvais état
e bosque alterado
غابة سيئة الحالة

865 domestic discharge
f rejet ménager
e desechos domésticos
تصريف النفايات المنزلية

866 domestic economy
f système économique local
e economía interna
اقتصاد محلي

867 domestic legislation
f législation interne
e legislación nacional
تشريع محلي

868 domestic needs
f besoins intérieurs
e necesidades internas
احتياجات محلية

869 domestic production
f production nationale
e producción nacional
انتاج محلي

870 domestic sewage
f eau usée d'origine ménagère
e aguas residuales domésticas
المجاري المنزلية

871 domestic solid wastes
f ordures ménagères
e residuos domésticos sólidos
نفايات منزلية صلبة

872 donor organism
f organisme donneur
e organismo donante
كائن متبرع

873 dose equivalent
f équivalent de dose
e equivalente de dosis
مكافئ الجرعة

874 dose-effect curve
f courbe dose-effet
e curva dosis-efecto
منحنى اثر الجرعة

875 dose-response functions
f fonctions doses-réactions
e funciones dosis-reacción
دالات الاستجابة للجرعة

876 dose-response relationship
f relation dose-réponse
e relación dosis-reacción
الصلة بين الجرعة والاستجابة

877 double cropping
f double culture
e doble cultivo
زراعة محصولين فيالسنة

878 downdraught
f courant descendant
e corriente descendente
تيار هوائي هابط

879 downstream course
f avalaison
e migración aguas abajo
باتجاه مجرى النهر

880 downstream ecosystem
f écosystème d'aval
e ecosistema de aguas abajo
نظام ايكولوجي سفلى

881 downward ozone transport
f transport descendant de l'ozone
e transporte descendente del ozono
نقل هابط للاوزون

882 downward spiral of poverty
f cercle vicieux de la pauvreté
e espiral descendente de la pobreza
الحلقة المفرغة للفقر

883 downward transport in the soil
f migration descendante dans le sous-sol
e movimiento descendente en el subsuelo
نقل هابط في التربة

884 downwash
f rabattement par la cheminée
e lavado a corriente descendente
انجراف سفلي

885 downwelling
f plongée d'eau
e corriente sumergente
غور (المياه السطحية)

886 downwind distance
f distance sous le vent
e distancia a sotavento
مسافة باتجاه الرياح

887 downwind source of pollution
f source de pollution sous le vent
e fuente de contaminación a sotavento
مصدر تلويث باتجاه الرياح

888 dredgings
f boues de dragage
e fango de dragado
طمي الاعماق

889 driftnet fishing
f pêche au filet dérivant
e pesca de enmalle y de deriva
صيد بشباك الجر

890 drinking water standard
f norme de qualité de l'eau potable
e norma de calidad del agua potable
معيار مياه الشرب

891 driving cycle
f cycle d'essai
e ciclo de pruebas
دورة تشغيل

892 droplet separator
f séparateur de gouttes
e separador de gotículas
جهاز فصل القطيرات

893 drought early warning
f avis précoce de sécheresse
e alerta anticipado de sequía
الانذار المبكر بالجفاف

894 **drought preparedness**
f prévention des situations de sécheresse
e preparación para casos de sequía
التأهب للجفاف

895 **drought probability map**
f carte des risques de sécheresse
e mapa de probabilidades de sequía
خريطة احتمالات الجفاف

896 **drought relief**
f secours en cas de sécheresse
e socorro en caso de sequía
الاغاثة من الجفاف

897 **drought resistant variety**
f variété résistante à la sécheresse
e variedad resistente a la sequía
صنف مقاوم للجفاف

898 **drought-tolerant species**
f végétation xérophile
e especies xerófilas
انواع تتحمل الجفاف

899 **drought watch**
f surveillance continue de la sécheresse
e vigilancia de la sequía
رصد الجفاف

900 **drought-prone region**
f région sujette à la sécheresse
e región expuesta a la sequía
منطقة معرضة للجفاف

901 **drum**
f fût
e tonel
برميل

902 **drumming**
f enfûtage
e envasado en toneles
تعبئة في براميل

903 **dry chemical**
f produit chimique sec
e producto químico seco
مادة كيميائية جافة

904 **dry cleaning solvent**
f solvant de nettoyage à sec
e solvente para limpieza en seco
مذيب للتنظيف الجاف

905 **dry deposition**
f dépôt (s) sec (s)
e deposición en seco
ترسب جاف

906 **dry grassland community**
f population d'herbacées xérophiles
e comunidad de pastos xerófilos
ماشية مراعي الكلأ الجفافي

907 **dry ice**
f neige carboniqe
e hielo seco
جليد جاف

908 **dry land**
f terre sèche
e tierra seca
ارض جافة

909 dry skin
f dessèchement de la peau
e piel seca
بشرة جافة

910 dry weather condition
f sécheresse
e tiempo seco
حالة طقس جاف

911 dry-land cultivation
f arido-culture
e cultivo de secano
زراعة الاراضي الجافة

912 dry-land rehabilitation
f remise en état de terres arides
e rehabilitación de tierras secas
اصلاح الارض الجافة

913 dump site
f dépotoir
e vertedero abierto
مدفن قمامة

914 dumping (of hazardous wastes)
f mise en décharge (de déchets dangereux)
e vertimiento(de desechos peligrosos)
القاء (النفايات الخطرة)

915 dumping at sea
f immersion en mer
e vertimiento en el mar
اغراق في البحر

916 dumping of toxic waste
f rejet de déchets toxiques
e vertimiento de desechos tóxicos
القاء النفايات السمية

917 dumping site
f lieu d'immersion
e lugar de vertimiento
موقع اغراق

918 dung
f déjections animales
e estiércol
روث

919 dust bowl
f région dénudée
e región de gran sequía
منطقة شبه صحراوية

920 dust deposit
f dépôts de poussières
e depósito de polvo
ترسب الغبار

921 dust discharge
f rejet de poussières
e descarga de polvo
تصريف الغبار

922 dustfall
f dépôts de poussières
e lluvia de polvo
ترسبات الغبار

923 dustiness
f pulvérulence
e pulverulencia
الغبرة

924 duty to re-import
f obligation de réimporter
e obligación de reimportar
واجب اعادة الاستيراد

925 dystrophic
f dystrophe
e distrófico
ناقص التغذية

926 dystrophic lake
f lac dystrophe
e lago distrófico
بحيرة قليلة المغذيات

E

927 earthly-environment
f environnement terrestre
e medio ambiente terrestre
بيئة ارضية

928 earthquake engineering
f génie parasismique
e tecnología antisísmica
هندسة الزلازل

929 earthquakes
f tremblements de terre
e terremotos
زلازل

930 easterlies
f vents d'est
e vientos del este
الرياح الشرقية

931 ecoagriculture
f écoagriculture
e ecoagricultura
الزراعة الايكولوجية

932 ecolabel
f écoétiquette
e etiqueta ecológica
علامة ايكولوجية

933 ecorisk
f risque écologique
e riesgo ecológico
مخاطرة ايكولوجية

934 ecotax
f écotaxe
e impuesto ecológico
ضريبة بيئية

935 ecotourism
f écotourisme
e turismo ecológico
سياحة بيئية

936 ecocide
f écocide
e ecocidio
ابادة ايكولوجية

937 ecoclimate
f écoclimat
e ecoclima
مناخ ايكولوجي

938 ecodevelopment
f écodéveloppement
e ecodesarrollo
تنمية ايكولوجية

939 ecological
f écologique
e ecológico
ايكولوجي

940 ecological balance
f équilibre écologique
e equilibrio ecológico
توازن ايكولوجي

941 ecological crimes
f délinquance écologique
e delitos ecológicos
جرائم ايكولوجية

942 ecological damage
f dommage écologique
e daño ecológico
ضرر ايكولوجي

943 ecological deficit
f déficit écologique
e déficit ecológico
عجز ايكولوجي

944 ecological disaster area
f zone de catastrophe écologique
e zona de desastre ecológico
منطقة كارثة ايكولوجية

945 ecological disruption
f perturbation écologique
e perturbación ecológica
خلل ايكولوجي

946 ecological efficiency
f efficacité écologique
e eficiencia ecológica
فعالية ايكولوجية

947 ecological factor
f facteur écologique
e factor ecológico
عامل ايكولوجي

948 ecological niche
f niche écologique
e nicho ecológico
مكمن ايكولوجي

949 ecological pyramid
f pyramide écologique
e pirámide ecológica
الهرم الايكولوجي

950 ecological rehabilitation
f restauration écologique
e rehabilitación ecológica
الاصلاح الايكولوجي

951 ecological security
f prévention des risques (écologiques)
e seguridad ecológica
الامن الايكولوجي

952 ecological suitability
f adaptabilité au milieu
e conveniencia desde el punto de vista ecológico
ملاءمة ايكولوجية

953 ecological sustainability
f durabilité écologique
e sostenibilidad ecológica
استدامة ايكولوجية

954 ecologically rational waste management
f gestion écologiquement rationnelle des déchets
e gestión ecológicamente racional de desechos
الادارة الرشيدة بيئيا للنفايات

955 ecologist
f écologue
e ecólogo
اخصائي بيئى

956 economic development
f développement économique
e desarrollo económico
تنمية اقتصادية

957 economic dualism
f dualisme économique
e dualismo económico
ازدواجية اقتصادية

958 economic feasability
f viabilité économique
e viabilidad económica
جدوى اقتصادية

959 economic management instruments
f instruments de gestion économique
e instrumentos de gestión económica
أدوات الإدارة الاقتصادية

960 economic planning
f planification économique
e planificación económica
تخطيط اقتصادى

961 economic zoning
f zonage économique
e zonificación económica
تقسيم المناطق اقتصاديا

962 economics of sustainability
f économie de la durabilité
e economía de la sostenibilidad
اقتصاد الاستدامة

963 ecoregion
f écorégion
e ecorregión
اقليم ايكولوجي

964 ecosphere
f écosphère
e ecosfera
المحيط الايكولوجي

965 ecosystem
f écosystème
e ecosistema
نظام ايكولوجي

966 ecosystem approach
f approche écosystémique
e enfoque ecosistémico
نهج النظام الايكولوجي

967 ecosystem diversity
f diversité des écosystèmes
e diversidad de ecosistemas
تنوع النظم الايكولوجية

968 ecosystem dysfunction
f dysfonction d'un écosystème
e disfunción de un ecosistema
إختلال فى النظام الايكولوجى

969 ecosystem integrity
f intégrité des écosystèmes
e integridad de los ecosistemas
سلامة النظام الايكولوجي

970 ecosystem rehabilitation
f restauration des écosystèmes
e rehabilitación de ecosistemas
اصلاح النظام الايكولوجي

971 ecotone
f écotone
e ecotono
منطقة تماس النظم الايكولوجية

972 **ecotoxicology**
f écotoxicologie
e ecotoxicología
علم السموم الايكولوجية

973 **ecotype**
f écotype
e ecotipo
نوع ايكولوجي

974 **eddy**
f tourbillon
e remolino
دوامة

975 **eddy diffusion**
f diffusion turbulente
e difusión turbulenta
انتشار دوامي

976 **edge effect**
f effet de lisière
e efecto de borde
اثر الحد

977 **eel grass**
f zostères
e zostera
عشبة الانقليس

978 **effect-oriented**
f pragmatique
e pragmático
موجه نحو الاثر

979 **efficiency of use of forest resources**
f valorisation des ressources forestières
e utilización eficiente de los recursos forestales
فعالية استخدام موارد الغابات

980 **effluent discharge**
f rejet d'effluents
e descarga de efluentes
تصريف الفضلات السائلة

981 **effluent standard**
f norme de rejet
e normas en materia de efluentes
معيار الفضلات

982 **egg-laying**
f ponte
e desove
وضع البيض

983 **egg-pod**
f oothèque
e ooteca
مجموعة البيض

984 **electric power**
f énergies électriques
e energía eléctrica
قوى كهربائية

985 **electric power distribution**
f distribution de l'énergie électrique
e distribución de energía eléctrica
توزيع القوى الكهربائية

986 electric power plants
f centrales électriques
e plantas de energía eléctrica
محطات توليد كهرباء

987 electrical engineering
f génie électrique
e ingeniería eléctrica
هندسة كهربائية

988 electrical storage devices
f procédés de stockage de l'électricité
e dispositivos para almacenar energía
أجهزة تخزين كهربائية

989 electrodialysis
f électrodialyse
e electrodiálisis
الفرز الغشائي الكهربائي

990 electromembrane technologies
f techniques électromembranaires
e tecnologías de electromembrana
تكنولوجيات الاغشية الكهربائية

991 electronic information network
f réseaux d'information électronique
e red electrónica de información
شبكة معلومات الكترونية

992 electrostatic precipitator
f dépoussiéreur électrostatique
e precipitador electrostático
جهاز ترسيب الكتروستاتي

993 elemental chlorine
f chlore élémentaire
e cloro elemental
عنصر الكلور

994 elemental sulphur
f soufre élémentaire
e azufre elemental
عنصر الكبريت

995 elimination
f élimination totale
e eliminación total
القضاء التام

996 embryo-based intervention
f intervention au niveau de l'embryon
e intervención a nivel embrionario
تدخل عند مرحلة الجنين

997 embryology
f embryologie
e embriología
علم الجنة

998 embryonic stem cell
f cellule souche de l'embryon
e célula embrionaria primaria
خلية ساقية جنينية

999 embryotechnology
f techniques d'intervention sur l'embryon
e embriotecnología
تكنولوجيا الاجنة

1000 **embryotransfer**
f transfert embryonnaire
e transferencia de embriones
نقل الاجنة

1001 **emergency area**
f zone sinistrée
e zona de emergencia
منطقة طوارئ

1002 **emergency assistance**
f assistance d'urgence
e ayuda de emergencia
مساعدة طارئة

1003 **emergency goods**
f biens de première nécessité
e artículos de socorro
بضائع لحالات الطوارئ

1004 **emergency relief**
f secours d'urgence
e socorro de emergencia
اغاثة في حالات الطوارئ

1005 **emergency relief measures**
f secours d'urgence
e socorro de emergencias
تدابير الإغاثة العاجلة

1006 **emergency shelter**
f abris de secours
e refugios de emergencia
مأوى عاجل

1007 **emergency stockpiles**
f stocks de secours d'urgence
e existencia de artículos de socorro
مخزونات الطوارئ

1008 **emerging crops**
f premières pousses
e cultivos incipientes
محاصيل ناشئة

1009 **emission**
f émission
e emisión
انبعاثات

1010 **emission behaviour**
f résultats quant à la réduction des émissions
e comportamiento de las emisiones
سلوك الانبعاثات

1011 **emission certification**
f homologation concernant les émissions
e certificación de las emisiones
ترخيص الانبعاثات

1012 **emission charges**
f redevances d'émission
e cargos por emisión
رسوم الانبعاثات

1013 **emission concentration**
f teneur à l'émission
e concentración de la emisión
تركيز الانبعاثات

1014 **emission control**
f lutte contre les émissions
e lucha contra las emisiones
مكافحة الانبعاثات

1015 emission factor
f coefficient d'émission
e coeficiente de emisión
عامل الانبعاثات

1016 emission load
f quantité (de polluants) émise
e cantidad (de contaminantes) emitida
كمية الانبعاثات

1017 emission performance
f résultats d'émission
e características de la emisión
اداء الانبعاثات

1018 emission point
f point d'émission
e punto de emisión
نقطة الانبعاثات

1019 emission rate
f débit d'émission
e tasa de emisión
معدل الانبعاث

1020 emission standard
f norme d'émission
e norma de emisión
معيار الانبعاث

1021 emission status
f état des émissions
e situación de las emisiones
حالة الانبعاثات

1022 emission strength
f teneur du rejet en polluant
e concentración de las emisiones
قوة الانبعاث

1023 emission-time pattern
f profil émission-temps
e perfil cronológico de las emisiones
نمط الانبعاثات – الوقت

1024 enabling environment
f environnement favorable
e medio ambiente favorable
بيئة مواتية

1025 enclosed sea
f mer fermée
e mar cerrado
بحر مغلق

1026 encroaching settlements
f habitat envahissant
e asentamientos invasores
مستوطنات زاحفة

1027 end-of-line technology
f technique de fin de chaîne
e tecnología de última etapa
تكنولوجيا المكافحة عند المصب

1028 end-of-pipe test
f essai au point de rejet
e prueba de etapa final
فحص عند المصب

1029 end-of-pipe treatment system
f système de traitement en aval
e sistema de tratamiento de etapa final
نظام معالجة عند المصب

1030 end-user
f utilisateur final
e usuario final
مستعمل نهائي

1031 end-user price
f prix payé par l'utilisateur final
e precio para el usuario final
السعر للمستعمل النهائي

1032 endangered animal species
f espèces animales en danger
e especies animales en peligro de extinción
أنواع الحيوانات المهددة بالانقراض

1033 endangered plant species
f espèces végétales en danger
e especies vegetales en peligro de extinción
أنواع النباتات المهددة بالانقراض

1034 endangered population
f population en grand danger
e población en peligro
أصناف مهددة

1035 endangered species
f espèce (s) menacée(s) d'extinction
e especies en peligro
انواع مهددة بالانقراض

1036 endemic
f endémique
e endémico
مستوطن

1037 endocrinology
f endocrinologie
e endocrinología
علم الغدد الصماء

1038 endothermic reaction
f réaction endothermique
e reacción endotérmica
تفاعل مستهلك للحرارة

1039 energy
f énergie
e energía
طاقة

1040 energy balance
f bilan énergétique
e balance energético
رصيد الطاقة

1041 energy budget
f bilan énergétique budgeté
e balance energético estimado
ميزانية الطاقة

1042 energy chain
f utilisation de l'énergie en cascade
e cadena energética
سلسلة الطاقة

1043 energy conservation
f économies d'énergie
e conservación de energía
صيانة الطاقة

1044 energy content
f énergie interne
e contenido energético
كمية الطاقة

1045 energy conversion efficiency
f rendement de conversion énergétique
e rendimiento conversión de la energía
فعالية تحويل الطاقة

1046 energy crop
f culture énergétique
e cultivo energético
محصول يستخدم لانتاج الطاقة

1047 energy economy
f économie de l'énergie
e economía de energía
علم اقتصاد الطاقة

1048 energy efficiency
f rendement énergétique efficient
e uso eficiente de la energía
فعالية الطاقة

1049 energy efficiency ratio
f taux de rendement énergétique
e coeficiente de rendimiento energético
نسبة كفاءة الطاقة

1050 energy input
f intrant énergétique
e insumo de energía
مدخلات الطاقة

1051 energy intensity
f intensité énergétique
e intensidad energética
كثافة الطاقة

1052 energy ladder
f échelle des énergies
e escala energética
سلم الطاقة

1053 energy output
f extrant énergétique
e producción de energía
مخرجات الطاقة

1054 energy policy
f politique énergétique
e política energética
سياسة الطاقة

1055 energy processes
f processus énergétiques
e proceso de energía
عمليات الطاقة

1056 energy production
f production d'énergie
e producción de energía
إنتاج الطاقة

1057 energy recovery from waste
f valorisation énergétique des déchets
e aprovechamiento energético de los desechos
استخلاص الطاقة من النفايات

1058 energy resources
f ressources énergétiques
e recursos energéticos
موارد الطاقة

1059 energy sink
f puits d'énergie
e sumidero de la energía
بالوعة طاقة

1060 energy sources
f sources d'énergie
e fuentes de energía
مصادر الطاقة

1061 energy strategy
f stratégie énergétique
e estrategia energética
استراتيجية الطاقة

1062 energy transition
f transition énergétique
e transición energética
تحول في مجال الطاقة

1063 energy use
f utilisation d'énergie
e utilización de la energía
استخدام الطاقة

1064 energy utilization patterns
f modèles d'utilisation énergétique
e modelos de uso de la energía
أنماط استخدام الطاقة

1065 energy wood
f bois énergie
e leña
حطب

1066 energy-efficient technology
f technologie à rendement énergétique élevé
e tecnología que usa la energía eficientemente
تكنولوجيا فعالة من حيث الطاقة

1067 energy-environment account
f compte de l'énergie et de l'environnement
e cuenta de la energía y el medio ambiente
حساب الطاقة والبيئة

1068 energy-intensive
f énergivore
e de alto consumo energético
كثيف الاستخدام للطاقة

1069 energy-saving
f économiseur d'énergie
e que ahorra energía
الاقتصاد في استهلاك الطاقة

1070 enforceable standard
f norme coercitive
e norma coercitiva
معيار قابل للتنفيذ

1071 engine design
f conception du moteur
e diseño del motor
تصميم المحرك

1072 engine durability
f longévité du moteur
e durabilidad del motor
عمر المحرك

1073 engineered adaptation
f adaptation conçue par l'homme
e adaptación técnica
تكييف مخطط

1074 engineered micro-organism
f micro-organisme modifié
e microorganismo manipulado
كائن مجهري محور

1075 engineering cost
f coût technique
e costo técnico
تكلفة الهندسة

1076 enriched uranium
f uranium enrichi
e uranio enriquecido
يورانيوم مخصب

1077 enteric fermentation
f fermentation entérique
e fermentación entérica
تخمر معوي

1078 entry into the environment
f pénétration dans l'environnement
e entrada en el medio ambiente
دخول في البيئة

1079 environment
f environnement
e medio ambiente
بيئة

1080 environment adequate for the health and well-being of individuals
f environnement propre à assurer la santé et le bien-être de chacun
e medio ambiente adecuado para la salud y el bienestar de las personas
بيئة مناسبة لصحة ورفاه الافراد

1081 environment adviser
f conseiller en environnement
e asesor ambiental
مستشار بيئي

1082 environment media
f compartiments de l'environnement
e sectores del medio ambiente
اوساط البيئة

1083 environment-damaging
f préjudiciable à l'environnement
e perjudicial para el medio ambiente
ضار بالبيئة

1084 environment-related disease
f maladie causée par l'environnement
e enfermedad relacionada con el medio ambiente
مرض بيئي

1085 environment-sensitive goods
f biens d'environnement
e productos compatibles con el medio ambiente
بضائع حساسة للبيئة

1086 environmental enhancement
f amélioration de l'environnement
e mejoramiento del medio ambiente
تحسين بيئي

1087 environmental image
f image de marque écologique
e imagen ambiental
صورة بيئية

1088 environmental impact statement
f notice d'impact sur l'environnement
e declaración sobre el impacto ecológico
بيان الأثر البيئى

1089 environmental impairment liability
f responsabilité de la dégradation de l'environnement
e responsabilidad por daños al medio ambiente
مسؤولية افساد البيئة

1090 environmental regulations
f réglementation de l'environnement
e reglamentación del medio ambiente
انظمة بيئية

1091 environmental release
f rejet dans l'environnement
e descarga en el medio ambiente
اطلاق في البيئة

1092 environmental abuses
f dégradation de l'environnement
e abuso del medio ambiente
اساءات بيئية

1093 environmental accident
f accident écologique
e accidente ecológico
حادث بيئي

1094 environmental accounting
f comptabilité de l'environnement
e contabilidad ambiental
محاسبة بيئية

1095 environmental alterant
f altéragène
e alterador del medio ambiente
مبدل بيئي

1096 environmental appraisal
f évaluation de l'environnement
e evaluación ambiental
تقييم بيئى

1097 environmental aspects of human settlements
f aspects écologiques des établissements humains
e aspectos ambientales de los asentamientos humanos
الجوانب البيئية للمستوطنات البشرية

1098 environmental assessment
f évaluation de l'environnement
e evaluación del medio ambiente
تقييم بيئى

1099 environmental assets
f patrimoine naturel
e patrimonio ambiental
اصول بيئية

1100 environmental attributes
f caractéristiques de l'environnement
e características del medio ambiente
صفات بيئية

1101 environmental audit(ing)
f audit écologique
e auditoría ambiental
مراجعة الحساب البيئي

1102 environmental awareness
f sensibilisation à l'environnement
e conciencia ambiental
وعي بيئي

1103 environmental behaviour
f comportement dans l'environnement
e comportamiento ambiental
سلوك بيئي

1104 environmental chemistry
f chimie de l'environnement
e química del medio ambiente
الكيمياء البيئية

1105 environmental concentration
f teneur dans l'environnement
e concentración en el medio ambiente
تركز بيئي

1106 environmental concern
f préoccupation écologique
e preocupación ecológica
قلق بيئي

1107 environmental conditions
f conditions ambiantes
e condiciones ambientales
ظروف بيئية

1108 environmental conservation
f protection de l'environnement
e conservación del medio ambiente
صيانة البيئة

1109 environmental constraints
f pressions sur l'environnement
e limitaciones ambientales
قيود بيئية

1110 environmental contingency planning
f plan d'intervention en cas de catastrophe
e planificación para contingencias ecológicas
تخطيط لحالات الطوارئ البيئية

1111 environmental costs
f coûts pour l'environnement
e costos para el medio ambiente
تكاليف بيئية

1112 environmental crimes
f crimes écologiques
e delitos ecológicos
جرائم بيئية

1113 environmental criteria
f critère écologiques
e criterios ecológicos
معايير بيئية

1115 environmental damage
f dégâts causés à l'environnement
e daños al medio ambiente
ضرر بيئي

1116 environmental data base
f base de données sur l'environnement
e base de datos ambientales
قاعدة بيانات بيئية

1117 environmental demand
f impératif écologique
e exigencia ambiental
مطلب بيئي

1118 environmental depletion
f appauvrissement de l'environnement
e empobrecimiento del medio ambiente
فقر البيئة

1119 environmental disaster
f catastrophe écologique
e desastre ecológico
كارثة بيئية

1120 environmental disorder
f pertubation de l'environnement
e perturbación del medio ambiente
خلل بيئي

1121 environmental dispute
f différend écologique
e controversia relativa al medio ambiente
نزاع بيئي

1122 environmental economic issues
f résultats économiques de l'écologie
e resultados económicos ambientales
قضايا اقتصادية بيئية

1123 environmental economics
f économie de l'environnement
e economía ambiental
اقتصاد البيئة

1124 environmental education
f éducation écologique
e educación ecológica
تثقيف بيئي

1125 environmental effectiveness
f efficacité en termes d'environnement
e eficacia ambiental
فعالية بيئية

1126 environmental emergency
f éco-urgence
e situación de emergencia relativa al medio ambiente
حالة طوارئ بيئية

1127 environmental emission
f émission dans l'environnement
e emisión en el medio ambiente
انبعثات في البيئة

1128 environmental engineer
f ingénieur écologue
e ingeniero ecólogo
مهندس بيئي

1129 environmental engineering
f génie écologique
e ingeniería ecológica
هندسة بيئية

1130 environmental epidemiology
f épidémiologie environnementale
e epidemiología ambiental
علم الامراض البيئية

1131 environmental ethics
f éthique de l'environnement
e ética del medio ambiente
قواعد السلوك البيئي

1132 environmental exposure
f exposition de l'environnement
e exposición del medio ambiente
تعرض بيئي

1133 environmental externalities
f effets sur l'environnement
e efectos en el medio ambiente
الآثار البيئية الخارجية

1134 environmental factor
f facteur environnemental
e factor ecológico
عامل بيئي

1135 environmental follow-up
f suivi écologique
e seguimiento ecológico
متابعة بيئية

1136 environmental forecasting
f prévision écologique
e pronóstico ecológico
تنبؤ بيئي

1137 environmental gains
f avantages du point de vue de l'environnement
e progreso ecológico
مكاسب بيئية

1138 environmental geology
f écogéologie
e geología ambiental
جيولوجيا بيئية

1139 environmental goods
f biens d'environnement
e bienes ambientales
منافع بيئية

1140 environmental hazard
f menace écologique
e peligro para el medio ambiente
خطر بيئي

1141 environmental health
f hygiène de l'environnement
e higiene ambiental
صحة بيئية

1142 environmental health impact assessment
f évaluation de l'impact de l'environnement sur la santé
e evaluación del impacto ambiental sobre la salud
تقييم أثر الصحة البيئية

1143 environmental health hazards
f risques écologiques pour la santé
e riesgos ambientales para la salud
مخاطر صحية بيئية

1144 environmental heritage
f patrimoine naturel
e patrimonio ambiental
تراث بيئي

1145 environmental impact
f impact sur l'environnement
e impacto sobre la ecología
أثر بيئى

1146 environmental impact assessment
f évaluation de l'impact sur l'environnement
e evaluación del impacto ambiental
تقييم الاثر البيئي

1147 environmental incentives
f motivations écologiques
e incentivos ecológicos
حوافز بيئية

1148 environmental indicator
f indicateur d'environnement
e indicador ecológico
مؤشر بيئي

1149 environmental industry
f écoindustrie
e industria del medio ambiente
صناعة بيئية

1150 environmental information
f information écologique
e información ecológica
معلومات بيئية

1151 environmental labelling
f éco-étiquetage
e etiquetado con indicaciones ecológicas
وضع العلامات البيئية

1152 environmental law
f droit de l'environnement
e derecho ambiental
قانون بيئي

1153 environmental levy
f taxe écologique
e impuesto ambiental
ضريبة بيئية

1154 environmental liability
f responsabilités en matière écologique
e responsabilidad ecológica
مسؤولية بيئية

1155 environmental limit concentration
f teneur limite dans l'environnement
e concentración ambiental límite
الحد البيئي للتركيزات

1156 environmental literacy
f conscience de l'environnement
e educación ecológica elemental
تعليم بيئي

1157 environmental management
f gestion de l'environnement
e ordenación del medio ambiente
ادارة بيئية

1158 environmental management indicators
f indicateurs de la gestion écologique
e indicadores de la gestión del medio ambiente
مؤشرات الإدارة البيئية

1159 environmental measures
f mesures de protection de l'environnement
e medidas de proteción ambiental
تدابير بيئية

1160 environmental microbiology
f écomicrobiologie
e microbiología ambiental
علم الجراثيم البيئية

1161 environmental misconduct
f méfaits écologiques
e conducta ecológica
سوء تصرف بيئى

1162 environmental monitoring
f surveillance de l'environnement
e vigilancia ambiental
رصد بيئي

1163 environmental monitoring satellite
f satellite de surveillance de l'environnement
e satélite de vigilancia ambiental
ساتل الرصد البيئي

1164 environmental movement
f mouvement écologiste
e movimiento ambientalista
حركة بيئية

1165 environmental norms and standards
f normes écologiques
e normas ecológicas
قواعد ومعايير بيئية

1166 environmental nuisance
f nuisances
e molestia ambiental
ازعاج بيئي

1167 environmental performance
f performance environnementale
e comportamiento ecológico
اداء بيئي

1168 environmental planning
f planification écologique
e planificación ambiental
تخطيط بيئي

1169 environmental policy
f politique de l'environnement
e política del medio ambiente
سياسة بيئية

1170 environmental pollution
f pollution
e contaminación ambiental
تلوث بيئي

1171 environmental product profile
f profil environnemental de produit
e perfil ambiental de los productos
موجز بيئي عن المنتجات

1172 environmental protection
f protection de l'environnement
e protección ambiental
حماية البيئة

1173 environmental quality
f qualité de l'environnement
e calidad del medio ambiente
نوعية البيئة

1174 environmental quality indicators
f indicateurs de la qualité de l'environnement
e indicadores de la calidad ambiental
مؤشرات النوعية البيئية

1175 environmental quality standard
f norme de qualité de l'environnement
e norma de calidad del medio ambiente
معيار نوعية البيئة

1176 environmental refugee
f réfugié écologique
e refugiado ecológico
لاجئ بيئي

1177 environmental renovation
f réhabilitation de l'environnement
e rehabilitación del medio ambiente
تجديد البيئة

1178 environmental requirements
f impératifs écologiques
e exigencias ambientales
متطلبات البيئة

1179 environmental resistance
f résistance du milieu
e resistencia del medio ambiente
مقاومة بيئية

1180 environmental resource base
f base de ressources écologiques
e base de recursos ambientales
قاعدة موارد بيئية

1181 environmental resources
f ressources de l'environnement
e recursos ecológicos
موارد بيئية

1182 environmental restoration
f restauration de l'environnement
e restauración del medio ambiente
استعادة البيئة

1183 environmental restructuring
f réaménagement du milieu
e reestructuración del medio ambiente
اعادة تشكيل البيئة

1184 environmental risk assessment
f évaluation des risques écologiques
e evaluación de riesgos ecológicos
تقييم المخاطر البيئية

1185 environmental safeguards
f mesures de protection de l'environnement
e medidas de protección ambiental
ضمانات بيئية

1186 environmental safety
f innocuité pour l'environnement
e inocuidad para el medio ambiente
سلامة بيئية

1187 environmental safety assessment
f évaluation des risques pour l'environnement
e evaluación de la inocuidad para el medio ambiente
تقييم السلامة البيئية

1188 environmental sanitation
f assainissement
e saneamiento del medio ambiente
المرافق الصحية البيئية

1189 environmental science
f science de l'environnement
e ciencia del medio ambiente
علم البيئة

1190 environmental security
f sécurité écologogique
e seguridad ecológica
الأمن البيئي

1191 environmental sensor
f capteur de variable (s) d'environnement
e sensor ambiental
جهاز استشعار بيئي

1192 environmental shock
f choc écologique
e choque ecológico
صدمة بيئية

1193 environmental soundness
f respect de l'environnement
e racionalidad ambiental
سلامة بيئية

1194 environmental spillover
f débordement environnemental
e extralimitación ambiental
تجاوز الحد البيئي

1195 environmental standard
f norme écologique
e norma ecológica
معيار بيئي

1196 environmental statistics
f statistiques de l'environnement
e estadísticas ambientales
أحصائيات بيئية

1197 environmental stock exchange
f bourse des valeurs écologiques
e bolsa ecológica de los valores
بورصة بيئية

1198 environmental stress
f stress environnemental
e tensión ambiental
اجهاد بيئي

1199 environmental stressor
f agresseur environnemental
e factor de tensión ambiental
عامل اجهاد بيئي

1200 environmental subsidies
f subventions écologiques
e subsidios ecológicos
إعانات بيئية

1201 environmental sustainability
f durabilité du point de vue de l'environnement
e durabilidad ambiental
الاستدامة البيئية

1202 environmental tax
f redevance environnementale
e impuesto ambiental
ضريبة بيئية

1203 environmental technology
f écotechnologie
e tecnología ambiental
تكنولوجيا بيئية

1204 environmental terminology
f terminologie écologique
e terminología ecológica
مصطلحات بيئية

1205 environmental threat
f menace pour l'environnement
e amenaza para el medio ambiente
تهديد بيئي

1206 environmental toxicity
f écotoxicité
e toxicidad para el medio ambiente
سمية بيئية

1207 environmental toxicology
f écotoxicologie
e toxicología ambiental
علم السموم البيئية

1208 environmental training
f formation à l'environnement
e capacitación ambiental
تدريب بيئي

1209 environmental valuation
f expertises écologiques
e peritaje ecológico
تقييم بيئي

1210 environmental vandalism
f vandalisme écologique
e vandalismo ecológico
تخريب بيئى

1211 environmental warfare
f guerre contre l'environnement
e guerra contra el medio ambiente
حرب بيئية

1212 environmental warning services
f services d'alerte-environnement
e servicios de alerta ambiental
خدمات الانذار البيئي

1213 environmental-friendly
f écophile
e inocuo para el medio ambiente
ملائم للبيئة

1214 environmentalist
f écologiste
e ambientalista
اخصائي بيئي

1215 environmentally compatible technologies
f techniques compatibles avec l'environnement
e tecnologías compatibles con el medio ambiente
تكنولوجيات ملائمة بيئيا

1216 environmentally acceptable substitute
f substitut écologiquement acceptable
e substituto ambientalmente aceptable
بديل مقبول بيئيا

1217 environmentally burdensome
f dommageable pour l'environnement
e perjudicial para el medio ambiente
عبء بيئي

1218 environmentally conscious
f sensibilisé à l'environnement
e con conciencia ambiental
واع بيئيا

1219 environmentally fragile area
f zone écologiquement fragile
e zona ecológicamente vulnerable
منطقة هشة بيئيا

1220 environmentally neglected zone
f zone écologiquement délaissée
e zona ecológicamente descuidada
منطقة مهملة بيئيا

1221 environmentally related diseases
f maladies liées à l'environnement
e enfermedades relacionadas con el medio ambiente
أمراض متصلة بالبيئة

1222 environmentally safe and sound technologies
f techniques sûres et écologiques
e tecnologías ecológicamente inocuas y racionales
التكنولوجيات المأمونة والسليمة بيئيا

1223 environmentally sensitive area
f région écologiquement sensible
e zona ecológicamente vulnerable
منطقة حساسة بيئيا

1224 environmentally sensitive sector
f secteur intégrant l'environnement
e sector ecológicamente vulnerable
قطاع حساس للبيئة

1225 environmentally sound
f écologiquement équilibré
e ecológicamente racional
سليم بيئيا

1226 environmentally sound and sustainable development
f dévéloppement durable et écologiquement rationnel
e desarrollo ecológicamente racional y sostenible
تنمية سليمة ومستدامة بيئيا

1227 environmentally sound and sustainable practices
f pratiques écologiquement rationnelles et viables
e prácticas ecológicamente racionales y sostenibles
ممارسات سليمة ومستدامة بيئيا

1228 environmentally sound characterisitics
f caractéristiques écologiques
e características ecológicamente adecuadas
خصائص سليمة بيئيا

1229 environmentally sound development
f écodévéloppement
e desarrollo ecológicamente racional
تنمية سليمة بيئيا

1230 environmentally sound energy system
f système énergétique écologiquement rationnel
e sistema energético ecológicamente racional
نظام طاقة سليم بيئيا

1231 environmentally sound management of inland water
f gestion écologiquement rationnelle des eaux intérieures
e ordenación ecológicamente racional de las aguas interiores
ادارة سليمة بيئيا للمياه الداخلية

1232 environmentally sound product
f produit écologique
e producto ecológicamente adecuado
منتج سليم بيئيا

1233 environmentally sound technological process
f procédé écotechnologique
e proceso tecnológico ecológicamente racional
عملية تكنولوجية سليمة بيئيا

1234 environmentally stressed
f précaire du point de vue de l'environnement
e sometido a tensiones ambientales
مجهد بيئيا

1235 environmentally sustainable development
f développement écologiquement durable
e desarrollo ecológicamente sostenible
تنمية مستدامة بيئيا

1236 environmentally unsound technology
technique anti-écologique
e tecnología ecológicamente inadecuada
تكنولوجيا غير سليمة بيئيا

1237 enzymes
f enzymes
e enzimas
أنزيمات

1238 epidemics
f épidémies
e epidemias
أوبئة

1239 epidemiology
f épidémiologie
e epidemiología
علم الأوبئة

1240 equatorial rain forest
f forêt ombrophile équatoriale
e selva pluvial ecuatorial
غابة مطيرة استوائية

1241 ergonomics
f ergonomie
e ergonomía
دراسة أحوال العمال وفاعليتهم

1242 estuarine biology
f biologie estuarienne
e biología de los estuarios
علم أحياء مصبات الأنهار

1243 estuarine conservation areas
f zones de conservation estuaires
e zonas estuarinas de conservación
مناطق صيانة مصبات الأنهار

1244 estuarine ecosystems
f écosystèmes estuariens
e ecosistemas estuarinos
النظم الايكولوجية لمصبات الأنهار

1245 estuarine oceanography
f océanographie estuarienne
e oceanografía estuarina
أقيانوغرافيا مصبات الأنهار

1246 eukaryotic organism
f eucaryote
e eucariota
كائن وحيد الخلية

1247 eutrophic
f eutrophe
e eutrófico
مغذ

1248 eutrophic lake
f lac eutrophe
e lago eutrófico
بحيرة غنية بالمغذيات

1249 eutrophication
f eutrophisation
e eutroficación
تخثث

1250 evaporation
f évaporation
e evaporación
تبخر

1251 evaporative emission
f émission par évaporation
e emisión por evaporación
انبعاث تبخري

1252 excessive use of fertilizer
f emploi abusif d'engrais
e utilización excesiva de fertilizantes
استخدام مفرط للاسمدة

1253 exclusion
f expérience d'exclusion
e experimento de exclusión
تجربة استبعادية

1254 exempt quantities
f quantités exemptées
e cantidades exentas
كميات معفاة

1255 exhaust emission
f émissions (de gaz) d'échappement
e emisiones de gases de escape
انبعاثات العادم

1256 exhaust emission limit
f limite d'émissions d'échappement
e límite de emisiones de gases de escape
حد انبعاثات العادم

1257 exhaust gas recirculation
f recyclage des gaz d'échappement
e recirculación de gases de escape
اعادة تدوير غاز العادم

1258 exhaust heat
f rejet thermique
e calor de escape
حرارة العادم

1259 exhaustion
f épuisement final
e agotamiento
استنفاد

1260 exosphere
f exosphère
e exosfera
الغلاف الجوي الخارجي

1261 exothermic reaction
f réaction exothermique
e reacción exotérmica
تفاعل مطلق للحرارة

1262 explant
f explant
e explante
غرز خارجي

1263 explosive
f matière explosive
e explosivo
متفجر

1264 export of hazardous wastes
f exportation de déchets dangereux
e exportación de desechos peligrosos
تصدير نفايات خطرة

1265 export state
f etat d'exportation
e estado exportador
دولة تصدير

1266 exposed
f exposé
e expuesto
معرض

1267 exposed area
f région vulnérable
e zona vulnerable
منطقة معرضة

1268 exposed crowns of trees
f partie ensoleillée de la cime des arbres
e copas expuestas de los árboles
قمم الاشجار المعرضة (للشمس)

1269 exposed to air
f au contact de l'air
e expuesto al aire
معرض للهواء

1270 exposure limit
f valeur limite d'exposition
e límite de exposición
حدود التعرض

1271 exposure pathway
f mode d'exposition
e trayectoria de exposición
طريق التعرض

1272 exposure time
f durée d'exposition
e tiempo de exposición
زمن التعرض

1273 ex situ conservation
f conservation ex situ
e conservación ex situ
حفظ خارج الموقع

1274 extended measurements
f mesures poussées
e mediciones ampliadas
قياسات موسعة

1275 extinct species
f espèce(s) disparue (s)
e especies extinguidas
انواع منقرضة

1276 extinguishing agent
f agent d'extinction
e agente extinctor
عامل اطفاء

1277 extraction of timber
f exploitation du bois
e extracción de madera
استغلال الخشب

1278 exudate
f sécrétion végétale
e exudación
افراز

1279 eye damage
f affections oculaires
e lesión oftálmica
ضرر بصري

1280 eye disorders
f troubles de la vue
e afecciones oftálmicas
اختلالات بصرية

1281 eyespot
f piétin-verse
e cercosporelosis
تبقع عيني

F

1282 fabric filter
f filtre en tissu
e filtro de tejido
مصفاة نسيجية

1283 facilitator
f animateur
e facilitador
ميسر

1284 factor of safety
f coefficient de sécurité
e coeficiente de seguridad
عامل السلامة

1285 factor use
f consommation de facteurs de production
e uso de los factores
استخدام عوامل النتاج

1286 faecal coliform bacteria
f bactérie fécale coliforme
e bacteria coliforme fecal
بكتيريا قولونية غائطية

1287 fallow cycle
f période de mise en jachère
e período de barbecho
دورة زراعية

1288 family planning
f planification familiale
e planificación de la familia
تنظيم الأسرة

1289 famine early warning
f alerte précoce à la pénurie alimentaire
e alerta de hambruna
انذار مبكر بالمجاعة

1290 fanning
f panache étalé
e dispersión en abanico horizontal
انتشار أفقي

1291 fanning plume
f panache horizontal
e penacho en abanico horizontal
عمود دخان أفقي

1292 farm buildings
f bâtiments agricoles
e edificios agrícolas
مبانى زراعية

1293 farm survey
f enquête agricole
e encuesta agrícola
استقصاء زراعي

1294 farmers rights
f droits de l'agriculteur
e derechos del agricultor
حقوق المزارعين

1295 farming sustainability
f durabilité de l'activité agricole
e sostenibilidad de la agricultura
استدامة الزراعة

1296 farmland
f terre agricole
e tierras de labranza
ارض زراعية

1297 farmyard waste
f déchets de ferme
e desechos de corrales
نفايات المزارع

1298 fast burn engine
f moteur à combustion rapide
e motor de combustión rápida
محرك سريع الاحتراق

1299 fast-growing tree
f arbre à croissance rapide
e árbol de crecimiento rápido
شجرة سريعة النمو

1300 fault
f faille
e falla
صدع

1301 fauna
f faune
e fauna
الحياة الحيوانية

1302 feed composition
f composition alimentaire pour animaux
e composición de los piensos animales
تركيب العلف

1303 feed conversion ratio
f indice de consommation
e coeficiente de conversión del alimento
نسبة تحول العلف

1304 feeder reservoir
f reservoire d'alimentation
e embalse de alimentación
خزان رافد

1305 feedlot
f élevage hors-sol
e corral de engorde
مرعي صغير

1306 feedstock
f produit de départ
e materia prima
مواد أولية

1307 fees deposit refund
f caution remboursable
e reembolso de derechos
رد عربون الرسوم

1308 felling cycle
f cycle d'abattage
e ciclo de corta
دورة قطع الاشجار

1309 fen
f marais
e marjal
مستنقع

1310 fermentable
f fermentescible
e fermentable
قابل للتخمير

1311 fiber reinforced plastic
f plastique renforcé par des fibres
e plástico reforzado con fibras
لدائن مقواة بالالياف

1312 field aggregator
f coefficient d'agrégation sur le champ
e coeficiente de agregación del campo
معامل التراكم في الحقل

1313 field capacity
f capacité au champ
e capacidad de campo
السعة الحلقية

1314 field crop
f culture de plein champ
e cultivo extensivo
محصول حقلي

1315 field genebank
f banque de gènes de terrain
e banco de genes sobre el terreno
مصرف جينات ميداني

1316 field-tested
f soumis à des essais sur le terrain
e ensayado sobre el terreno
مجرب ميدانيا

1317 filling
f comblement
e relleno
ردم

1318 filter (layer) separator
f séparateur filtrant
e filtro separador
طبقة فاصلة مرشحة

1319 filters
f filtres
e filtros
مرشاح

1320 filtration rate
f vitesse de filtration
e velocidad de filtración
معدل الترشيح

1321 financial assistance
f aide financière
e asistencia financiera
مساعدة مالية

1322 fine textured soil
f sol de particules fines
e suelo fino
تربة ناعمة

1323 fire safety requirements
f règles de prévention des incendies
e medidas de seguridad contra incendios
متطلبات الوقاية من الحرائق

1324 first-order consumers
f consommateurs de premier ordre
e consumidores primarios
مستهلكون من الدرجة الاولى

1325 fish
f poissons
e peces
أسماك

1326 fish culture
f pisciculture
e piscicultura
تربية الأسماك

1327 fish nursery
f alevinage
e criadero de peces
مفرخة اسماك

1328 fish shoal
f banc de poissons
e banco de peces
سرب اسماك

1329 fish stock
f stock de poissons
e población de peces
الرصيد السمكي

1330 fisheries biology
f biologie halieutique
e biología pesquera
بيولوجيا الاسماك

1331 fisheries management
f gestion de pêcheries
e ordenación de la pesca
إدارة مصايد الأسماك

1332 fishery body
f organe régional chargé des pêches
e organismo encargado de la pesca
هيئة مصائد اسماك

1333 fishing vessels
f bateaux de pêche
e buques pesqueros
سفن صيد الأسماك

1334 fishmeal
f farine de poisson
e harina de pescado
مسحوق سمكى

1335 fixed ammonia
f ammoniac fixe
e amoníaco fijo
امونيا مثبتة

1336 fixed combustion source
f foyer fixe
e foco fijo de combustión
مصدر احتراق ثابت

1337 fixed gas
f gaz difficilement liquéfiable
e gas difícilmente licuable
غاز ثابت

1338 fixed property
f propriété immuable
e característica inmutable
خاصية ثابتة

1339 fixed protection standard
f norme de protection fixe
e norma de protección fija
معيار حماية ثابت

1340 flammable liquids
f liquides inflammables
e líquidos inflamables
سوائل قابلة للاشتعال

1341 flash flood
f crue soudaine
e crecida repentina
فيضان سريع

1342 flat-rate reduction
f réduction (des émissions) à un taux uniforme
e reducción a una tasa uniforme
تخفيض موحد

1343 flatlands
f basses plaines
e planicies
سهول

1344 flexible foam
f mousse souple
e espuma
رغوة مرنة

1345 floc
f floc
e flóculo
كتلة متلبدة

1346 flocculation
f floculation
e floculación
تلبد

1347 flood control
f lutte anti-inondation
e control de inundaciones
مكافحة الفيضان

1348 flood control system
f système de protection contre les inondations
e sistema de defensa contra las inundaciones
نظام التحكم بالفيضانات

1349 flood forecasting
f prévision des crues
e previsión de inundaciones
تنبؤ بالفيضانات

1350 flood hazard
f risque de crue
e riesgo de inundación
خطر الفيضان

1351 flood level
f cote d'alerte d'inondation
e nivel de la inundación
مستوى الفيضان

1352 flood mitigation
f atténuation des inondations
e reducción de las inundaciones
تخفيف حدة الفيضان

1353 flood mitigator
f écrêteur de crues
e medio de control de inundaciones
مخفف حدة الفيضان

1354 flood pattern
f régime des crues
e régimen de inundaciones
نمط الفيضان

1355 flood protection
f protection contre les crues
e protección contra las inundaciones
حماية من الفيضانات

1356 flood warning
f annonce des crues
e alerta de inundaciones
انذار بالفيضان

1357 floods
f inondations
e inundaciones
فيضانات

1358 flora
f flore
e flora
الحياة النباتية

1359 flow area
f section d'écoulement
e sección de flujo
منطقة تدفق

1360 flow filter
f filtre traversant
e filtro de flujo
مصفاة تدفق

1361 flow rate
f débit
e caudal
معدل الدفق

1362 flue (gas) dust
f cendres volantes
e polvo de combustión
غبار (غاز) المداخن

1363 flue gas
f gaz de cheminée
e gas de chimenea
غاز المداخن

1364 flue gas reheating
f réchauffage par les gaz de cheminés
e recalentamiento por los gases de chimenea
اعادة تسخين غاز المداخن

1365 fluidized bed combustion
f combustion en lit fluidisé
e combustión en lecho fluidizado
احتراق على قاعدة مميعة

1366 fluorination
f fluoration
e fluoración
فلورة

1367 fly ash
f cendres volantes
e cenizas volátiles
رماد متطاير

1368 flyway
f voies de migration
e vías migratorias
مسارات هجرة

1369 foam blowing
f gonflement de la mousse
e espumación
ارغاء

1370 fog
f brouillard
e niebla y bruma
ضباب

1371 fogging
f brumisage
e nebulización
تكوين الضباب

1372 foliage injury
f lésion foliaire
e daños causados al follaje
ضرر باوراق النبات

1373 foliage monitoring
f suivi de l'évolution du feuillage
e observación sistemática del follaje
رصد اوراق الاشجار

1374 foliar feed
f engrais foliaire
e alimento foliar
علف ورقي

1375 follow-up testing
f essais ultérieurs
e ensayos complementarios
اختبار متابعة

1376 food
f aliments
e alimentos
اغذية

1377 food (bearing) tree
f arbre donnant des fruits comestibles
e árbol que da alimentos
شجرة مثمرة

1378 food additives
f additifs alimentaires
e aditivos alimenticios
مضافات الأغذية

1379 food chain
f chaîne trophique
e cadena trófica
سلسلة غذائية

1380 food colourants
f colorants alimentaires
e colorantes alimenticios
ملونات الأغذية

1381 food contamination
f contamination des denrées alimentaires
e contaminación de alimentos
تلوث الأغذية

1382 food intake
f prise de nourriture
e consumo de alimentos
إستهلاك غذائى

1383 food irradiation
f irradiation des aliments
e irradiación de alimentos
تعرض الأغذية للإشعاع

1384 food pattern
f régime alimentaire
e régimen alimentario
نمط غذائي

1385 food poisoning
f intoxication alimentaire
e intoxicación alimentaria
تسمم غذائي

1386 food preservation
f conservation des aliments
e preservación de alimentos
حفظ الأغذية

1387 food safety
f sécurité alimentaire
e seguridad alimentaria
سلامة غذائية

1388 food science
f science alimentaire
e ciencia de los alimentos
علم الأغذية

1388 food science
f science alimentaire
e ciencia de los alimentos
علم الأغذية

1389 food storage
f stockages des aliments
e almacenamiento de alimentos
تخزين الأغذية

1390 food technology
f technologie alimentaire
e tecnología alimenticia
تكنولوجيا الأغذية

1391 food transport
f transport des aliments
e transporte de alimentos
نقل الأغذية

1392 food web
f réseau trophique
e red alimentaria
شبكة غذائية

1393 forest area
f zone forestière
e zona de bosques
منطقة غابات

1394 forest biomass
f biomasse forestière
e biomasa forestal
كتلة احيائية حرجية

1395 forest clearing
f déboisement
e tala del bosque
ازالة الغابات

1396 forest conservation
f préservation des forêts
e conservación forestal
صيانة الغابات

1397 forest cover
f couverture forestière
e cubierta forestal
غطاء حرجي

1398 forest crown
f houppiers cimes des arbres de la forêt
e copas de los árboles
قمم اشجار الغابات

1399 forest decay
f dévitalisation des forêts
e decadencia forestal
اضمحلال الغابات

1400 forest depletion
f destruction de la forêt
e agotamiento forestal
استنفاد الغابات

1401 forest deterioration
f dépérissement des forêts
e deterioro forestal
تدهور الغابات

1402 forest dieback
f dépérissement (terminal) des forêts
e extinción paulatina del bosque
موت تدريجي للغابات

1403 forest ecology
f écologie forestière
e ecología forestal
ايكولوجيا الغابات

1404 forest edibles
f produits comestibles de la forêt
e productos forestales comestibles
منتجات حرجية غذائية

1405 forest environment
f milieu forestier
e medio ambiente forestal
البيئة الحرجية

1406 forest estates
f propriété forestière
e propiedad forestal
أملاك حرجية

1407 forest fallow
f jachères forestières
e barbecho forestal
أرض حرجية بور

1408 forest fire
f incendie de forêt
e incendio forestal
حريق الغابات

1409 forest floor
f couverture
e tapiz vegetal
ارضية الغابة

1410 forest health
f santé des forêts
e salud de los bosques
صحة الغابات

1411 forest heritage
f patrimoine forestier
e patrimonio forestal
تراث حرجي

1412 forest industries
f industries du bois
e industrias forestales
صناعات حرجية

1413 forest land
f espace forestier
e superficie forestal
ارض حرجية

1414 forest legal instrument
f instrument relatif aux forêts
e instrumento relativo a los bosques
صك حرجي

1415 forest loss
f pertes forestières
e pérdida de bosques
خسارة الغابات

1416 forest management
f aménagement forestier
e ordenación de los bosques
ادارة الغابات

1417 forest manager
f aménagiste des forêts
e técnico forestal
مدير حراج

1418 forest policy
f politique forestière
e políticas forestales
سياسة الغابات

1419 forest products
f produits forestiers
e productos forestales
منتجات حرجية

1420 forest range
f massif forestier
e bosques
سلسلة جبال حرجية

1421 forest regeneration
f régénération des forêts
e regeneración del bosque
تجديد الحراج

1422 forest replantation
f repeuplement des forêts
e replante de bosques
اعادة التحريج

1423 forest resource assessment
f inventaire des ressources forestières
e evaluación de recursos forestales
تقييم موارد الغابات

1424 forest seed orchard
f verger à graines forestières
e huerto de semillas forestales
مستنبت اشجار حرجية

1425 forest soil
f sol forestier
e suelo forestal
تربة الغابات

1426 forest species
f essence (forestière)
e especie forestal
انواع حرجية

1427 forest strategy
f stratégie d'aménagement forestier
e plan de ordenación forestal
استراتيجية الغابات

1428 forest taxation
f imposition de la forêt
e impuesto forestal
ضريبة لحفظ الغابات

1429 forester
f sylviculteur
e silvicultor
مستغل الغابات

1430 forestry
f sylviculture
e silvicultura
حراجة

1431 forestry legislation
f législation forestière
e legislación forestal
تشريع حراجى

1432 forestry management
f aménagement forestier
e silvicultura (gestión)
ادارة الحراج

1433 fossil (fuel) conversion technology
f techniques de conversion des combustibles fossiles
e tecnología de conversión de combustibles fósiles
تكنولوجيا تحويل الوقود الحفرى

1434 fossil fuel
f combustible fossile
e combustible fósil
وقود حفرى

1435 foundation nursery
f alevinier de départ
e vivero de partida
مشتل اساسي

1436 foundry cupola
f cubilot de fonderie
e cubilote de fundición
فرن الصهر

1437 fragile ecosystems
f écosystèmes fragiles
e ecosistemas delicados
نظم ايكولوجية هشة

1438 fragmentation
f morcellement
e fragmentación
تجزئة

1439 free rider
f bénéficiaire automatique
e beneficiario automático
منتفع

1440 free temperature rise
f hausse de température naturelle
e diferencia de temperatura
ارتفاع الحرارة الطبيعية

1441 freeze-dry
f lyophiliser
e liofilizar
جفف بالتجميد

1442 freshwater
f eau douce
e agua dulce
مياه عذبة

1443 freshwater acidification
f acidification des cours d'eau
e acidificación del agua dulce
زيادة حموضة المياه العذبة

1444 freshwater and marine aquaculture
f aquiculture d'eau douce et d'eau de mer
e acuicultura de agua dulce y de mar
تربية المائيات في المياه العذبة ومياه البحار

1445 freshwater biology
f biologie de l'eau douce
e biología del agua dulce
علم أحياء المياه العذبة

1446 freshwater degradation
f détérioration des eaux douces
e deterioro de las aguas dulces
تدهور المياه العذبة

1447 freshwater ecosystems
f écosystèmes d'eau douce
e ecosistemas de agua dulce
النظم الايكولوجية للمياه العذبة

1448 freshwater management
f gestion de l'eau douce
e ordenación de las aguas dulces
ادارة المياه العذبة

1449 freshwater monitoring
f surveillance des eaux douces
e vigilancia del agua dulce
رصد المياه العذبة

1450 freshwater prawn
f crevette d'eau douce
e camarón de agua dulce
اربيان المياه العذبة

1451 freshwater resources
f ressources en eau douce
e recursos de agua dulce
موارد المياه العذبة

1452 frost point
f point de gelée blanche
e punto de escarcha
نقطة الصقيع

1453 fuel alcohol
f alcool à usage de combustibles
e alcohol combustible
كحول الوقود

1454 fuel desulphurisation
f désulfuration des combustibes
e desulfuración de combustibles
ازالة الكبريت من الوقود

1455 fuel economy
f réduction de la consommation de carburant
e reducción del consumo de combustible
اقتصاد في استهلاك الوقود

1456 fuel metering
f dosage du carburant
e medición del combustible
قياس كمية الوقود

1457 fuel mixture
f mélange combustible
e mezcla combustible
مزيج وقودي

1458 fuel oil
f fioul
e fuelóleo
زيت الوقود

1459 fuel performance
f rendement du carburant
e rendimiento del combustible
اداء الوقود

1460 fuel quality standards
f normes de qualité des combustibles
e normas de calidad de los combustibles
معايير نوعية الوقود

1461 fuel substitution
f recours à des combustibles de substitution
e sustitución de combustibles
استبدال الوقود

1462 fuel switching away from oil
f renoncement aux hydrocarbures
e sustitución del petróleo por otros combustibles
التحول عن الوقود النفطي

1463 fuel-efficient cities
f villes consommant moins de combustibles
e ciudades de bajo consumo de combustible
مدن مقتصدة فى الوقود

1464 fuelwood
f bois de feu
e leña
حطب

1465 fuelwood plantation
f plantation d'essences pour bois de feu
e plantación de árboles para leña
مزرعة خشب الوقود

1466 fugitive emission
f émission fugace
e emisiones fugitivas
انبعاث هارب

1467 full-scale test
f essai en grandeur réelle
e ensayo a escala natural
اختبار شامل

1468 fumigant
f fumigant
e fumigante
مادة متبخرة مطهرة

1469 fumigation
f fumigation
e fumigación
تبخير

1470 fungi
f champignons
e hongos
فطريات

1471 fungicides
f fongicides
e fungicidas
مبيدات الفطر

G

1472 game fish
f tout-gros
e especies de pesca deportiva
سمك رياضة الصيد

1473 game management
f gestion du gibier
e regulación de la caza
ادارة حيوانات الصيد

1474 garbage collection
f enlèvement des ordures ménagères
e recolección de basura
جمع القمامة

1475 garbage grinding
f broyage d'ordures
e trituración de basura
طحن القمامة

1476 gas analyzer
f analyseur de gaz
e analizador de gases
محلل الغاز

1477 gas chromatography
f chromatographie en phase gazeuse
e cromatografía en fase gaseosa
فصل لونى للغاز

1478 gas cleaning
f épuration des gaz
e depuración de gases
تنقية الغاز

1479 gas liquefaction
f liquéfaction des gaz
e licuefacción de gases
إسالة الغاز

1480 gas scrubber
f laveur de gaz
e lavador de gas
جهاز تنقية الغاز

1481 gas-borne particles
f particules en suspension dans un gaz
e partículas en suspensión gaseosa
جسيمات عالقة بالغاز

1482 gas-to-particle conversion
f transformation du gaz en particules
e transformación de gas en partículas
تحويل الغاز الى جزيئات

1483 gaseous hydrocarbon
f hydrocabure gazeux
e hidrocarburo gaseoso
هيدروكربونات غازية

1484 gaseous pollutants
f polluants gazeux
e contaminantes gaseosos
ملوثات غازية

1485 gender issues
f questions se rapportant aux différences de sexe
e cuestiones de género
قضايا الجنسين

1486 gene banks
f banques de gènes
e banco de genes
مصارف الجينات

1487 gene mapping
f cartographie génique
e cartografía genómica
رسم الخرائط الجينية

1488 gene therapy
f thérapie génique
e terapia génica
علاج جيني

1489 gene transfer
f transfert de gènes
e transferencia de genes
نقل الجينات

1490 genet
f organisme présentant une originalité génétique
e organismo genéticamente diferente
كائن مختلف جينيا

1491 genetic biodiversity resources
f ressoures génétiques de la diversité biologique
e recursos genéticos de la diversidad biológica
موارد جينية للتنوع البيولوجي

1492 genetic conservation
f conservation génétique
e conservación genética
حفظ الجينات

1493 genetic diversity
f diversité génétique
e diversidad genética
تنوع جيني

1494 genetic engineering
f génie génétique
e ingeniería genética
هندسة جينية

1495 genetic erosion
f érosion génétique
e erosión genética
اضمحلال جيني

1496 genetic make-up
f constitution génétique
e constitución genética
تركيب جيني

1497 genetic trait
f caractère génétique
e característica genética
خاصية جينية

1498 genetically modified insect resistance
f résistance aux insectes induite par modification génétique
e resistencia a los insectos obtenida por modificaciones genéticas
مقاومة للحشرات محفزة بالتحوير الجيني

1499 genetically modified organism
f organisme génétiquement modifié
e organismo modificado genéticamente
كائن محور جينيا

1500 genetics
f génétique
e genética
جينات

1501 genotype
f génotype
e genotipo
نمط جيني

1502 genus
f genre
e género
جنس ؛ نوع

1503 geoengineering
f géo-ingénierie
e ingeniería geológica
الهندسة الجيولوجية

1504 geographic distribution of resources
f distribution géographique des ressources naturelles
e distribución geográfica de los recursos naturales
توزيع جغرافى للموارد

1505 geographic ozone distribution
f répartition spatiale de l'ozone
e distribución geográfica del ozono
توزع جغرافي للاوزون

1506 geographic variation
f variation spatiale
e variación geográfica
تباين جغرافي

1507 geology
f géologie
e geología
علم طبقات الارض

1508 geomorphology
f géomorphologie
e geomorfología
علم شكل الارض

1509 geophysics
f géophysique
e geofísica
علم الفيزياء الأرضية

1510 geoscience
f sciences de la terre
e ciencias geológicas
علم الارض

1511 geosphere
f géosphère
e geosfera
المحيط الارضى

1512 geotechnics
f géotechnique
e geotecnia
تقنية التربة

1513 geotechnology
f géotechnologie
e geotecnología
تكنولوجيا التربة

1514 geothermal energy
f énergie géothermique
e energía geotérmica
طاقة حرارية أرضية

1515 geothermal gradient
f gradient géothermique
e gradiente geotérmico
تدرج الحرارة الجوفية

1516 germ cell
f cellule germinale
e célula germinal
خلية جرثومية

1517 germ plasm
f germoplasmes (collections de semences)
e plasma germinal(colección de semillas)
بلازما جرثومية

1518 glacial phase
f phase glaciaire
e período glaciar
مرحلة جليدية

1519 glacial records
f archives glaciaires
e archivos glaciares
سجلات الجليديات

1520 glacial varve
f varve glaciaire
e varva glaciar
طبقات ترسبية جليدية

1521 glaciology
f glaciologie
e glaciología
علم الجليديات

1522 glass collection unit
f colonne de collecte sélective de verre
e unidad de separación de vidrio
وحدة جمع الزجاج

1523 global conventions
f conventions mondiales
e convenios globales
اتفاقيات عالمية

1524 global cycle
f cycle global
e ciclo global
دورة عالمية

1525 global emission
f émissions globales
e emisiones globales
انبعاثات عالمية

1526 global environment
f environnement planétaire
e medio ambiente mundial
بيئة عالمية

1527 global environmental problem
f problème mondial lié à l'environnement
e problema mundial del medio ambiente
مشكلة بيئية عالمية

1528 global ozone distribution
f répartition mondiale de l'ozone
e distribución mundial del ozono
التوزع العالمي للاوزون

1529 global ozone transport
f transport planétaire de l'ozone
e transporte mundial del ozono
النقل العالمي للاوزون

1530 global response
f parade mondiale
e reacción mundial
استجابة عالمية

1531 global security
f sécurité du globe
e seguridad mundial
الامن العالمي

1532 global stewardship
f bonne intendance globale
e administración ambiental mundial
الاشراف العالمي

1533 global warming
f réchauffement de la planète
e calentamiento de la tierra
الاحترار العالمي

1534 good housekeeping practices
f bonnes méthodes d'économie domestique
e buenos métodos de economía doméstica
ممارسات التدبير المنزلي الجيدة

1535 good laboratory practice
f bonne pratique de travail en laboratoire
e buenas prácticas de laboratorio
ممارسة مختبرية جيدة

1536 government buildings
f édifices publics
e edificios gubernamentales
مبانى حكومية

1537 government environmental expenditures
f dépenses écologiques gouvernementales
e gastos ecológicos estatales
النفقات البيئية الحكومية

1538 grade
f qualité
e calidad
درجة (نقاء)

1539 grade sample
f échantillon pris au hasard
e muestra tomada al azar
عينة عشوائية

1540 gradient
f gradient
e gradiente
تدرج

1541 grain fumigant
f fumigant pour céréales
e fumigante de cereales
داخنة الحبوب

1542 grain size analysis
f analyse granulométrique
e análisis granulométrico
تحليل حجم الحبة

1543 grains
f grains
e cereales
حبوب

1544 granular material
f granulats
e granulados
مادة حبيبية

1545 granulation
f granulation
e granulación
تكوين الحبيبات

1546 grass cover
f couvert herbacé
e cubierta herbácea
غطاء عشبي

1547 grass fires
f brûlage des pâturages
e quema de pastos
حرائق الاعشاب

1548 grassland
f prairies
e pradera
أرض معشوشبة

1549 grassland ecosystems
f écosystèmes des pâturages
e ecosistemas de pastizales
النظم الايكولوجية للمراعي

1550 grassroots efforts
f initiatives locales
e actividades a nivel popular
جهود شعبية

1551 gravel bed filter
f filtre à gravier
e filtro de grava
طبقة حصى مرشحة

1552 gravity separation
f séparation gravitaire
e separación por gravedad
فصل بالجاذبية

1553 grazed forest
f forêt pâturable
e bosque pastable
غابة مرعية

1554 grazing animal
f ruminant
e animal que pasta
حيوانات الرعي

1555 grazing land
f patûrages
e tierra de pastoreo
مرعى

1556 grease resistance
f résistance à l'huile et aux graisses
e resistencia a las grasas
مقاومة الشحوم

1557 grease tank
f bac à graisses
e tanque de grasas
صهريج الشحوم

1558 green belt
f ceinture verte
e cinturón verde
حزام اخضر

1559 green car
f voiture verte
e automóvil ecológico
سيارة خضراء

1560 green cover
f couvert végétal
e cubierta vegetal
غطاء اخضر

1561 green fiscal instruments
f instruments fiscaux écologiques
e instrumentos fiscales ecológicos
أدوات مالية بيئية

1562 green land
f surfaces en herbe
e tierra cubierta de vegetación
ارض خضراء

1563 greenhouse climate
f climat de serre
e clima de invernadero
مناخ الاحتباس الحرارى

1564 greenhouse effects
f effet(s) de serre
e efecto(s) [de] invernadero
آثار الاحتباس الحرارى

1565 greenhouse gas
f gaz à effet de serre
e gas de efecto invernadero
غاز الاحتباس الحرارى

1566 greenhouse gas sequestration
f retenue des gaz à effet de serre
e retención de los gases de efecto invernadero
تنحية ايونات غاز الاحتباس الحرارى

1567 greening of the economy
f écologisation de l'économie
e armonización de la economía con el medio ambiente
مواءمة بين الاقتصاد والبيئة

1568 greening of the world
f reverdissement des campagnes
e reverdecimiento del planeta
تحريج العالم

1669 ground clearance
f débroussaillement
e desbroce
تنظيف الارض

1570 ground cover
f couverture des sols
e cubierta del suelo
غطاء ارضى

1571 ground facilities
f moyens au sol
e instalaciones en tierra
مرافق ارضية

1572 ground station
f station au sol
e estación terrestre
محطة ارضية

1573 ground-based observation
f observation au sol
e observaciones hechas en tierra
مراقبة أرضية

1574 groundwater
f eau(x) souterraine(s)
e aguas subterráneas
مياه جوفية

1575 groundwater flow
f écoulement de la nappe d'eau souterraine
e flujo de aguas subterráneas
تدفق المياه الجوفية

1576 groundwater runoff
f ruissellement souterrain
e escorrentía de aguas subterráneas
جريان المياه الجوفية

1577 groundwater seepage
f infiltration dans la nappe phréatique
e infiltración en las aguas subterráneas
تسرب الى المياه الجوفية

1578 group right
f droit collectif
e derecho colectivo
حق جماعي

1579 growing season
f saison de croissance
e estación de crecimiento
فصل النمو

1580 growth (inhibition) test
f essai d'inhibition de la croissance
e ensayo de inhibición del crecimiento
اختبار (منع) النمو

1581 growth promoter
f activateur de croissance
e promotor del crecimiento
حافز النمو

1582 growth regulating gene
f gène régulateur de croissance
e gen regulador del crecimiento
جينة تنظيم النمو

1583 growth ring
f cerne annuel
e anillo de crecimiento (de un árbol)
حلقة نمو

H

1584 habitat
f habitat
e hábitat
موئل

1585 habitat conservation
f conservation des habitats
e conservación del hábitat
صيانة المو ائل

1586 habitat destruction
f destruction des habitats
e destrucción del hábitat
تدمير المو ائل

1587 habitat dislocation
f bouleversement des habitats
e trastorno del hábitat
اضطراب الموئل

1588 habitat management
f gestion et aménagement des habitats
e ordenación del hábitat
ادارة الموئل

1589 haematology
f hématologie
e hematología
علم الدم

1590 hailstorm
f chute de grêle
e tormenta de granizo
عاصفة برد

1591 halogen(ated) derivative
f dérivé-halogéné
e derivado halogenado
مشتق مهلجن

1592 halogenated
f halogéné
e halogenado
مهلجن

1593 halogenated extinguishing agent
f extincteur halogéné
e sustancia ignífuga halogenada
عامل اطفاء مهلجن

1594 hard coal
f houille
e antracita, hulla
الفحم القاسي

1595 hard water
f eau dure
e agua dura
ماء عسر

1596 harmful effect
f effet délétère
e efecto nocivo
اثر ضار

1597 harmful substance
f substance nocive
e sustancia nociva
مادة ضارة

1598 harvest of forest trees
f coupe des arbres forestiers
e explotación forestal
قطع اشجار الحراج

1599 harvesting
f récolte
e cosecha
صيد

1600 hazard
f danger
e peligro
خطر

1601 hazard area
f zone à risque
e zona de riesgo
منطقة خطر

1602 hazard assessment
f évaluation des dangers
e evaluación de la peligrosidad
تقييم الاخطار

1603 hazard classification system
f système de classement des dangers
e sistema de clasificación de los riesgos
نظام تصنيف الاخطار

1604 hazard concentration limit
f limite de concentration en substances dangereuses
e limite de concentración peligrosa
حد تركيز الخطر

1605 hazard mapping
f établissement de cartes de risques
e levantamiento de mapas de peligros
رسم خرائط مناطق الخطر

1606 hazard of pollutants
f dangers des polluants
e riesgos de los contaminantes
مخاطر الملوثات

1607 hazard proneness
f prédisposition aux risques
e propensión o exposición a riesgos
قابلية التعرض للخطر

1608 hazard rating
f évaluation du danger
e evaluación de los peligros
تقدير الخطر

1609 hazard-prone area
f zone à risque
e zona expuesta a riesgos
منطقة معرضة للخطر

1610 hazard-resisting building
f bâtiment à l'épreuve des risques
e edificio seguro
مبنى مقاوم للاخطار

1611 hazardous characteristics
f caractéristiques de danger
e características peligrosas
خصائص خطرة

1612 hazardous substance
f substance dangereuse
e sustancia peligrosa
مادة خطرة

1613 hazardous wastes
f déchets dangereux
e desechos peligrosos
النفايات الخطرة

1614 haze
f brume sèche
e bruma, neblina
اغبرار

1615 headstream
f affluent (du cours) supérieur
e afluente de cabecera
رافد (نهري) قرب المنبع

1616 health care
f soins de santé
e cuidado de la salud
رعاية صحية

1617 health facilities
f installations sanitaires
e centros de salud pública
مرافق صحية

1618 health hazard
f risque pour la santé
e peligro para la salud
خطر على الصحة

1619 health legislation
f législation sanitaire
e legislación sanitaria
تشريع صحى

1620 health risk assessment
f appréciation des risques pour la santé
e evaluación de los riesgos para la salud
تقدير الأخطار على الصحة

1621 health-related biotechnologies
f biotechnologies à visée sanitaire
e biotecnologías relacionadas con la sanidad
تكنولوجيات حيوية متعلقة بالصحة

1622 health-related monitoring
f surveillance à visée sanitaire
e vigilancia relacionada con la salud
الرصد المتعلق بالصحة

1623 healthy city
f ville saine
e ciudad sana
مدينة صحية

1624 healthy environment
f environnement sain
e medio ambiente sano
بيئة صحيحة

1625 heat balance
f bilan thermique
e balance térmico
توازن حراري

1626 heat capacity
f capacité calorifique
e capacidad calorífica
السعة الحرارية

1627 heat content
f enthalpie
e contenido calorífico
المحتوى الحراري

1628 heat engineering
f thermique
e termotecnia
الهندسة الحرارية

1629 heat factor
f facteur de réchauffement
e factor térmico
عامل الحرارة

1630 heat flow
f flux de chaleur
e flujo térmico
تدفق حراري

1631 heat insulation
f calorifugeage
e aislamiento térmico
عزل حراري

1632 heat load
f surcharge thermiqe
e carga térmica
حمل حراري

1633 heat resistance
f résistance à la chaleur
e resistencia al calor
مقاومة الحرارة

1634 heat sink
f puits de chaleur
e sumidero del calor
بالوعة حرارية

1635 heat stabilizing
f thermostabilisant
e estabilización térmica
مثبت للحرارة

1636 heat stress
f stress thermique
e estrés calórico
اجهاد حراري

1637 heat stroke
f coup de chaleur
e insolación
ضربة شمس

1638 heat-trapping ability
f capacité de rétention de la chaleur
e capacidad de retención de calor
قدرة على حبس الحرارة

1639 heat(lands)
f landes
e brezal
أراض بور

1640 heating plant
f installation de chauffage
e instalación de calefacción
منشأة تسخين

1641 heavy goods vehicle
f poids lourd
e vehículo de carga pesada
مركبة احمال ثقيلة

1642 heavy ice
f glace dense
e hielo espeso
جليد كثيف

1643 heavy metals
f métaux lourds
e metales pesados
فلزات ثقيلة

1644 heavy rain
f fortes pluies
e lluvia fuerte
مطر غزير

1645 heavy-duty engine
f moteur de grosse cylindrée
e motor de gran potencia
محرك قوي الاحتمال

1646 hedgerow
f haie
e seto
سياج

1647 herbicides
f herbicides
e herbicidas
مبيدات الأعشاب

1648 heritage coast
f littoral protégé
e litoral protegido
ساحل تراثى محمي

1649 heritage of humanity
f patrimoine de l'humanité
e patrimonio de la humanidad
تراث الانسانية

1650 heritage preservation
f préservation du patrimoine
e preservación del patrimonio
حفظ التراث

1651 heritage site
f site protégé
e sitio protegido
موقع محمي

1652 heterotroph
f hétérotrophe
e heterótrofo
متباين التغذية

1653 hidden hunger
f faim insoupçonnée
e hambre oculta
جوع مستتر

1654 hidden pollution
f pollution cachée
e contaminación oculta
تلوث مستتر

1655 high bogs
f hauts-marais
e pantanos de montaña
مستنقعات مرتفعة

1656 high protein foods
f aliments riches en protéines
e alimentos con alto contenido de proteínas
أغذية عالية البروتين

1657 high sea fisheries
f pêche hauturière
e pesca de altura
مصائد اسماك اعالي البحار

1658 high vacuum
f vide poussé
e alto vacío
تفريغ عالي

1659 high water
f pleine mer
e pleamar
المد

1660 high-capacity container
f caisson de grande capacité
e contenedor de gran capacidad
حاوية كبيرة

1661 high-rise buildings
f tours d'habitation
e edificios altos
مبانى عالية

1662 high-sulphur fuel
f combustible à haute teneur en soufre
e combustible de alto contenido de azufre
وقود عالي نسبة الكبريت

1663 higher productivity variety
f variété à meilleur rendement
e variedad de mayor rendimiento
صنف ذو انتاجية مرتفة

1664 highland ecosystems
f écosystèmes des hauts plateaux
e ecosistemas de los altiplanos
النظم الايكولوجية للمرتفعات

1665 highlands
f hauts plateaux
e tierras altas
المرتفعات

1666 highways
f autoroutes
e autopistas
طرق رئيسية

1667 historical sites
f sites historiques
e sitios históricos y monumentos
مواقع تاريخية

1668 holistic management of water
f gestion holistique de l'eau
e ordenación holística del agua
الادارة الكلية للمياه

1669 holistic overview
f aperçu général holistique
e visión holística
استعراض كلي

1670 homelessness
f personnes sans abris
e personas sin hogar
تشرد

1671 homestead gardens
f jardins des exploitations agricoles
e huertos particulares
بساتين منزلية

1672 homogeneous layer
f couche homogène
e capa homogénea
طبقة التجانس السطحي

1673 homolytic breaking
f rupture homolytique
e fragmentación homolítica
تجزؤ متجانس

1674 horizontal diffusion
f diffusion horizontale
e difusión horizontal
انتشار افقي

1675 horizontal transfer
f transfert horizontal
e transferencia horizontal
انتقال افقي

1676 **horticultural crops**
f cultures horticoles
e cultivos hortícolas
محاصيل البساتين

1677 **horticulture**
f horticulture
e horticultura
بستنة

1678 **hospital wastes**
f déchets des hôpitaux
e desechos hospitalarios
نفايات المستشفيات

1679 **hospitals**
f hôpitaux
e hospitales
مستشفيات

1680 **host organism**
f organisme-hôte
e organismo huésped
كائن عضوي مضيف

1681 **household refuse**
f ordures ménagères
e basuras (domésticas)
نفايات منزلية

1682 **housing density**
f densité de logements
e densidad de vivienda
كثافة الإسكان

1683 **housing finance**
f financement du logement
e financiación de la vivienda
تمويل الإسكان

1684 **housing improvements**
f amélioration des logements
e mejoras de vivienda
تحسينات فى الإسكان

1685 **housing legislation**
f législation immobilière
e legislación de vivienda
تشريع للإسكان

1686 **housing needs**
f besoins en logements
e necesidades de vivienda
احتياجات الإسكان

1687 **housing programmes**
f programmes de logements
e programas de vivienda
برامج الإسكان

1688 **housing quality standards**
f normes de qualité des logements
e normas de calidad de vivienda
معايير نوعية الإسكان

1689 **human biology**
f biologie humaine
e biología humana
علم الأحياء البشرى

1690 **human capital**
f capital humain
e capital humano
رأس المال البشري

1691 **human diseases**
f maladies humaines
e enfermedades humanas
أمراض بشرية

1692 human environment
f cadre de vie
e medio humano
البيئة البشرية

1693 human exposure
f exposition des personnes
e exposición del ser humano
تعرض بشري

1694 human exposure to pollutants
f exposition de l'homme aux polluants
e exposición humana a contaminantes
تعرض الإنسان للملوثات

1695 human health
f santé humaine
e salud humana
الصحة البشرية

1696 human migration
f migrations humaines
e migraciones humanas
هجرة بشرية

1697 human pathology
f pathologie humaine
e patología humana
علم الأمراض البشرية

1698 human physiology
f physiologie humaine
e fisiología humana
علم وظائف الأعضاء البشرية

1699 human population
f population humaine
e población humana
سكان

1700 human rights
f droits de l'homme
e derechos humanos
حقوق الإنسان

1701 human settlements
f établissements humains
e asentamientos humanos
المستوطنات البشرية

1702 human settlements management
f gestion des établissements humains
e administración de los asentamientos humanos
إدارة المستوطنات البشرية

1703 human welfare
f bien-être de l'humanité
e bienestar de la humanidad
رفاه الانسان

1704 human-induced degradation
f dégradation anthropique
e degradación debida a actividades humanas
تدهور بفعل الانسان

1705 human-made disasters
f catastrophes provoquées par l'homme
e desastres provocados por el hombre
كوارث من صنع الإنسان

1706 humectant
f humectant
e humectante
مادة مانعة للجفاف

1707 humid tropics
f zone tropicale humide
e zonas tropicales húmedas
المناطق المدارية الرطبة

1708 humidity
f humidité
e humedad
رطوبة

1709 hunting
f chasse
e caza
صيد

1710 hurricane
f ouragan
e huracán
اعصار مداري

1711 hurricane tracking
f détection des ouragans
e seguimiento de huracanes
تتبع الأعاصير

1712 hybridoma
f hybridome
e hibridoma
خلية التهجين

1713 hydraulic conductivity
f conductivité hydraulique
e conductividad hidráulica
النفاذية المائية

1714 hydraulic connection
f communication entre masses d'eau
e conexión hidráulica
اتصال مائي

1715 hydraulic gradient
f gradient hydraulique
e gradiente hidráulico
تدرج الضغط المائي

1716 hydraulic head
f charge hydraulique
e carga hidráulica
طاقة الضغط الهيدروليكي

1717 hydrocarbon compounds
f composés hydrocarburés
e compuestos de hidrocarburos
مركبات الهيدروكربونات

1718 hydrocarbons
f hydrocarbures
e hidrocarburos
هيدروكربونات

1719 hydrochloric acid
f acide chlorhydrique
e ocirdيhrolc odicل
حامض الهيدروكلوريك

1720 hydroclimate
f hydroclimat
e hidroclima
مناخ مائي

1721 hydroelectric power
f énergie hydro-électrique
e energía hidroeléctrica
قوى كهرمائية

1722 hydrographic survey
f levé hydrogaphique
e levantamiento hidrográfico
مسح مائي

1723 hydrographic water basin
f bassin hydrographique
e cuenca hidrográfica
حوض مائي

1724 hydrographical chart
f carte hydrographique
e mapa hidrográfico
خريطة مائية

1725 hydrography
f hydrographie
e hidrografía
علم المساحة

1726 hydrology
f hydrologie
e hidrología
علم المياه

1727 hydrometeorological survey
f étude hydrométéorologique
e estudio hidrometeorológico
دراسة الرطوبة الجوية

1728 hydromorphic soil
f sol hydromorphe
e suelo hidromórfico
تربة يتغير تكوينها بتعريضها للماء

1729 hydroponics
f cultures hydroponiques
e hidroponía
الزراعة فوق الماء

1730 hydrosphere
f hydrosphère
e hidrosfera
الغلاف المائي

1731 hygienic waste disposal facility
f système d'évacuation des déchets respectueux des normes d'hygiène
e sistema de eliminación higiénica de desechos
مرفق تصريف صحي للنفايات

1732 hygroscopic
f hygroscopique
e higroscópico
ممتص للرطوبة

I

1733 ice
f glaces
e hielo
جليد

1734 ice age
f époque glaciaire
e glaciación
عصر جليدي

1735 ice cap
f calotte glaciaire
e casquete glaciar
غطاء جليدي

1736 ice foot
f pied de glace
e pie de hielo
الرصيف الجيلدي

1737 ice shelf
f écueil de glace
e barrera de hielo
جرف جليدي

1738 ice-crystal chemistry
f chimie des cristaux de glace
e química de los cristales de hielo
كيمياء البلورات الجليدية

1739 identification mark
f marque d'identification
e marcas distintivas
علامة تعريف

1740 identification of pollutants
f identification des polluants
e identificación de contaminantes
تحديد الملوثات

1741 illegal tipping
f dépôts sauvages
e vertimiento ilegal de basura
إلقاء غير مشروع

1742 imaging sensor
f détecteur imageur
e sensor de formación de imágenes
جهاز استشعار تصويري

1743 imbibition
f imbibition
e imbibición
تشرب

1744 immediate action
f intervention immédiate
e medidas inmediatas
إجراء فوري

1745 immediate danger area
f zone de danger immédiat
e zona de peligro inmediato
منطقة خطر مباشر

1746 immission
f immission
e inmisión
التركيز عند مستوى سطح الأرض

1747 immune suppression
f immuno-suppression
e inmunosupresión
إعاقة المناعة

1748 immunological diseases
f maladies immunologiques
e enfermedades inmunológicas
أمراض المناعة المرضية

1749 immunology
f immunologie
e inmunología
علم المناعة

1750 impact analysis
f étude d'impact
e evaluacion de los efectos
تحليل التأثير

1751 impact statement
f notice d'impact
e exposición de los efectos
بيان الأثر (البيئي)

1752 impaction of particles
f retombées de particules
e impacto de partículas
تصادم الجزئيات

1753 impervious layer
f formation imperméable
e capa impermeable
طبقة غير منفذة

1754 impinger
f appareil à impact
e percutor
مرطم

1755 import state
f état d'importation
e estado de importación
دولة الاستيراد

1756 impoundment
f installation de confinement
e confinamiento
محتجز

1757 improved technologies
f perfectionnements techniques
e tecnologías (más) avanzadas
تكنولوجيات محسنة

1758 improved variety
f variété améliorée
e variedad mejorada
نوع محسن

1759 in vitro storage
f conservation in vitro
e almacenamiento in vitro
تخزين في أنابيب الاختبار

1760 in-plant atmosphere
f atmosphère d'usine
e atmósfera del interior de la fábrica
الجو داخل المصنع

1761 in-plant recycling
f recyclage sur place
e reciclaje interno
إعادة التدوير داخل المصنع

1762 inactive
f inactif
e inactivo
خامل

1763 inadvertent loss
f perte fortuite
e pérdida accidental
خسارة عرضية

1764 incentive
f incitation
e incentivo
حافز

1765 incentive schemes
f mécanismes d'incitation
e sistemas de incentivo
خطط الحفز

1766 incident report
f déclaration d'incident
e informe de incidentes
تقرير عن الحادث

1767 incidental catch
f prise accidentelle
e captura incidental
الصيد العارض

1768 incidental emission
f émission accidentelle
e emisión ocasional
انبعاث عارض

1769 incidental release
f rejet accidentel
e descarga accidental
إطلاق عارض

1770 incineration of waste
f incinération des déchets
e incineración de desechos
ترميد النفايات

1771 incineration plant
f usine d'incinération
e planta de incineración
محرقة

1772 incinerator
f incinérateur
e incinerador
جهاز حرق

1773 incipient inhibition
f début d'inhibition
e inhibición incipiente
منع اولى

1774 incoming radiation
f rayonnement incident
e radiación incidente
الأشعة الساقطة

1775 incremental cost
f surcoût
e costo adicional
تكلفة إضافية

1776 indicator organism
f organisme indicateur
e indicador ecológico
كائن دليلي

1777 indigenous forest
f forêt naturelle
e bosque autóctono
غابة أصلية

1778 indigenous people
f populations autochtones
e población autóctona
السكان الاصليون

1779 indigenous species
f espèces locales
e especie autóctona
الأنواع الأصلية

1780 indigenous vector
f vecteur indigène
e vector autóctono
ناقل اصلي

1781 indirect-use value
f valeur d'usage indirecte
e valor de uso indirecto
قيمة الاستعمال غير المباشر

1782 indiscriminate logging
f exploitation forestière sans discernement
e explotación forestal indiscriminada
قطع الأشجار العشوائي

1783 indiscriminate use of chemicals
f abus des produits chimiques
e abuso de productos químicos
إساءة استعمال المواد الكيميائية

1784 individual source
f source individuelle
e fuente individual
مصدر فردي

1785 indoor air pollutant
f polluant de l'air intérieur des locaux
e contaminante del aire en locales cerrados
ملوثات الهواء الداخلي

1786 indoor air pollution
f pollution atmosphérique intérieure
e contaminación del aire interior
تلوث الهواء الداخلي

1787 indoor air ventilation
f aération
e ventilación de locales cerrados
تهوية داخلية

1788 indoor climate
f conditions atmosphériques à l'intérieur des locaux
e condiciones atmosféricas en locales cerrados
الجو الداخلي

1789 indoor pollution
f pollution à l'intérieur des bâtiments
e contaminación en locales cerrados
التلويث الداخلي

1790 inducer
f inducteur
e inductor
مستحث

1791 industrial areas
f zones industrielles
e zonas industriales
مناطق صناعية

1792 industrial buildings
f bâtiments industriels
e edificios industriales
مباني صناعية

1793 industrial effluents
f effluents industriels
e efluentes industriales
ملوثات صناعية

1794 industrial emissions
f émissions industrielles
e emisiones industriales
انبعاثات صناعية

1795 industrial fumes
f fumées industrielles
e vapores industriales
أدخنة صناعية

1796 industrial legislation
f législation industrielle
e legislación industrial
تشريع صناعى

1797 industrial materials
f matériaux industriels
e materiales industriales
مواد صناعية

1789 industrial noise
f bruits industriels
e ruido industrial
ضوضاء صناعية

1799 industrial pattern
f structure industrielle
e características de la industria
نمط صناعي

1800 industrial processes
f processus industriels
e procesos industriales
عمليات صناعية

1801 industrial production statistics
f statistiques de la production industrielle
e estadísticas de la producción industrial
أحصائيات الإنتاج الصناعى

1802 industrial products
f produits industriels
e productos industriales
منتجات صناعية

1803 industrial wastewaters
f effluents industriels
e efluentes industriales
النفايات الصناعية السائلة

1804 industry
f industrie
e industria
الصناعة

1805 inert
f inerte
e inerte
خامل

1806 infant mortality
f mortalité infantile
e mortalidad infantil
وفاة الرضع

1807 infauna
f endofaune
e endofauna
حيوانات قاعيه

1808 infestation of crops
f infestation des récoltes
e infestación de las cosechas
إصابة المحاصيل

1809 infestation of food
f infestation des aliments
e infestación de alimentos y cosechas
إصابة الأغذية

1810 infiltration rate
f vitesse d'infiltration
e velocidad de infiltración
معدل التغلغل

1811 inflow of air
f afflux d'air
e afluencia de aire
تيار الهواء الداخل

1812 inflow pipe
f conduite d'amenée
e cañería de entrada
أنبوبة الدفق الداخل

1813 influencing factor
f facteur déterminant
e factor influyente
عامل مؤثر

1814 information infrastructure
f moyens d'information
e infraestructura de información
البنية الاساسية للمعلومات

1815 information networks
f réseaux d'informations
e redes de información
شبكات معلومات

1816 infrared-absorbing gas
f gaz absorbant le rayonnement infrarouge
e gas que absorbe rayos infrarrojos
غاز ماص للأشعة تحت الحمراء

1817 infrastructure
f infrastructure
e infraestructura
بنية أساسية

1818 injection well
f puits d'injection
e pozo de inyección
بئر دفن

1819 injury
f lésion
e lesión
إصابة

1820 inland fresh water system
f réseau hydrographique d'eau douce
e sistema de aguas dulces interiores
شبكة المياه العذبة الداخلية

1821 inland water bodies
f eaux intérieures
e masas de agua interiores
الأجسام المائية الداخلية

1822 inland water transport
f transport par voie de navigation intérieure
e transporte fluvial
نقل فى المياه الداخلية

1823 inland waters
f eaux intérieures
e aguas interiores
مياه داخلية

1824 inland waterways
f voies de navigation intérieures
e vías fluviales de navegación
طرق النقل المائى الداخلى

1825 inlet swirl
f turbulence à l'admission
e turbulencia de admisión
دوامة المدخل

1826 innocuous to the environment
f sans danger pour l'environnement
e inocuo para el medio ambiente
غير ضار بالبيئة

1827 inorganic chemistry
f chimie inorganique
e química inorgánica
كيمياء غير عضوية

1828 inorganic pollutants
f polluants inorganiques
e contaminantes inorgánicos
ملوثات غير عضوية

1829 inorganic substances
f substances inorganiques
e sustancias inorgánicas
مواد غير عضوية

1830 insect pest resistance
f résistance aux attaques d'insectes
e resistencia a los insectos
المقاومة للإصابة بالحشرات

1831 insert
f inséré
e inserción
وليجة

1832 inserted material
f matériau inséré
e material insertado
مادة مولجة

1833 instrument measurements
f mesures instrumentales
e mediciones con instrumentos
قياسات بأجهزة

1834 insulating foam
f mousse isolante
e espuma aislante
رغوة عازلة

1835 intake
f dose
e dosis
جرعة

1836 intake port
f conduite d'admission
e lumbrera de admisión
فتحة الدخول

1837 intangible
f bien incorporel
e bien intangible
غير محسوس

1838 integrated economic-environmental accounting
f comptabilité économique et environnementale intégrée
e contabilidad integrada de la economía y el medio ambiente
نظام متكامل للمحاسبة الاقتصادية – البيئية

1839 integrated environmental controls
f mesures intégrées de protection de l'environnement
e medidas integradas de protección del medio ambiente
ضوابط بيئية متكاملة

1840 integrated forest programme
f programme intégré sur les forêts
e programa forestal integrado
برنامج حرجي متكامل

1841 integrated monitoring
f surveillance intégrée
e vigilancia integrada
رصد متكامل

1842 integrated pest management
f lutte intégrée contre les ravageurs
e control integrado de las plagas
المكافحة المتكاملة للآفات

1843 intensive tree growing
f ligniculture
e silvicultura intensiva
غرس مكثف للأشجار

1844 inter-relations
f rapports réciproques
e interrelaciones
اوجه الترابط

1845 inter-tidal zone
f zone médiolittorale
e zona de intermareas
منطقة مدية

1846 interaction of pesticides
f interaction des pesticides
e interacción de plaguicidas
تفاعل مبيدات الآفات

1847 interference mechanism
f mécanisme d'interférence
e mecanismo de interferencia
آلية تدخل

1848 interfuel substitution
f substitution des combustibles
e sustitución de combustibles
إبدال الوقود

1849 intergenerational equity
f équité intergénérationnelle
e equidad entre generaciones
العدالة بين الأجيال

1850 interlaboratory test
f essai interlaboratoires
e ensayo entre laboratorios
اختبار مشترك بين المختبرات

1851 interlinkage
f articulation
e vinculación
ترابط

1852 intermediate species
f espèce active
e especie intermedia
أنواع وسيطة

1853 intermediate water
f eau intermédiaire
e agua intermedia
مياه وسيطة

1854 internal combustion engine
f moteur à combustion interne
e motor de combustión interna
محرك ذو احتراق داخلي

1855 internal crop marketing
f vente des produits agricoles dans le village
e venta local de productos agrícolas
تسويق المحاصيل في الداخل

1856 internal wave
f onde interne
e onda interna
موجة داخلية

1857 internalization
f internalisation
e internalización
تدخيل العوامل الخارجية

1858 internalization of environmental costs
f internalisation des coûts écologiques
e internalización de los costos ecológicos
تكاليف بيئية لتدخل العوامل الخارجية

1859 international environmental relations
f relations écologiques internationales
e relaciones ecológicas internacionales
العلاقات البيئية الدولية

1860 international river basins
f bassins fluviaux internationaux
e cuencas fluviales internacionales
أحواض الأنهار الدولية

1861 international standardization
f normalisation internationale
e normas internacionales
توحيد المقاييس الدولية

1862 international trade
f commerce international
e comercio internacional
تجارة دولية

1863 international watercourses
f cours d'eau internationaux
e cursos de agua internacionales
مجارى المياه الدولية

1864 internationally important ecosystems
f écosystèmes d'importance internationale
e ecosistemas de importancia internacional
نظم ايكولوجية مهمة دوليا

1865 interspecific hybridization
f hybridation entre espèces différentes
e hibridación interespecífica
التهجين بين الأنواع المختلفة

1866 intrazonal soil
f sol intrazonal
e suelo intrazonal
تربة شبه ناضجة

1867 invertebrates
f invertébrés
e invertebrados
لا فقريات

1868 invisible radiation
f rayonnement invisible
e radiación invisible
إشعاع غير منظور

1869 iodine deficiency disorders
f troubles dûs à une carence en iode
e trastornos por carencia de yodo
اضطرابات نقص اليود

1870 ion balance
f bilan ionique
e equilibrio iónico
توازن الأيون

1871 ion exchanger
f échangeur d'ions
e intercambiador de iones
مبادل ايوني

1872 ion source
f source d'ions
e fuente de iones
مصدر للأيونات

1873 ionization constant
f constante d'ionisation
e constante de ionización
ثابت التأين

1874 ionizing radiation
f radiation ionisante
e radiación ionizante
إشعاع مؤين

1875 ionosphere
f ionosphère
e ionosfera
الطبقة المتأنية

1876 iron industry
f industrie du fer
e industria del hierro
صناعة الحديد

1877 irradiance
f éclairement
e irradiancia
الإشعاعية

1878 irrecoverable loss
f perte irrémédiable
e pérdida irrecuperable
خسارة لا تعوض

1879 irrigation
f irrigation
e irrigación
ري

1880 irrigation canals
f canaux d'irrigation
e canales de irrigación
قنوات الري

1881 irrigation farming
f culture irriguée
e agricultura de irrigación
زراعة مروية

1882 irritant
f irritant (agent)
e sustancia irritante
مهيج

1883 island ecosystems
f écosystèmes insulaires
e ecosistemas insulares
النظم الايكولوجية للجزر

1884 isolated source
f source isolée
e fuente aislada
مصدر منعزل

1885 isolation of a pollutant
f isolement d'un polluant
e aislamiento de un contaminante
فصل مادة ملوثة

1886 isopsophic index
f indice isopsophique
e índice isopsófico
الرقم القياسي لضجيج الطائرات

1887 issue-wide goal
f objectif général
e objetivo general
هدف عام

J

1888 June drop
f coulure
e caída prematura de los frutos
سقوط الثمار قبل ينوعها

1889 juvenile fish
f juvéniles (de poissons)
e peces inmaturos
صغار السمك

K

1890 katabatic wind
f vent catabatique
e viento catabático
ريح هابطة

1891 killed lime
f chaux éteinte
e cal apagada
جير مطفأ

L

1892 laboratory-based experiment
f expérience en laboratoire
e experimento en laboratorio
تجربة مختبرية

1893 lacustrine sedimentology
f sédimentologie lacustre
e sedimentología lacustre
علم ترسبات البحيرات

1894 lag effect
f effet retardateur
e efecto de retardo
تأثير التخلف

1895 lagoon
f étang
e estanque
بحيرة شاطئية

1896 lake basins
f bassins lacustres
e cuencas lacustres
أحواض بحيرات

1897 lake ecosystem
f écosystème lacustre
e ecosistema lacustre
النظام الايكولوجي للبحيرات

1898 lakelands
f zones lacustres
e zonas lacustres
منطقة بحيرات

1899 lakes
f lacs
e lagos
بحيرات

1900 land allotment
f répartition des terres
e parcelación de la tierra
تخصيص الأرض

1901 land area of the globe
f surface émergée du globe
e superficie continental
مساحة اليابسة في الكرة الأرضية

1902 land capability analysis
f analyses agrométriques
e estudio del potencial de producción de las tierras
تحليل القدرات الزراعية للأراضي

1903 land care
f protection des terres
e protección de las tierras
العناية بالأراضي

1904 land carrying capacity
f capacités optimales du sol
e capacidad de sostenimiento de la tierra
قدرة حمل الارض

1905 land clearance
f défrichage
e desmonte
أرض مهيأة للزراعة

1906 land clearing
f défrichement
e desmonte
قطع الأشجار

1907 land degradation
f dévastation des terres
e degradación de tierras
تدهور الأرض

1908 land disposal site
f décharge
e vertedero
مدفن (نفايات) بري

1909 land erosion control
f lutte contre l'érosion des sols
e lucha contra la erosión de los suelos
مكافحة تآكل التربة

1910 land loss
f pertes en terres
e pérdida de tierras
خسارة الأراضي

1911 land mammals
f mammifères terrestres
e mamíferos terrestres
ثدييات أرضية

1912 land masses
f terres émergées
e masas continentales
كتل أرضية

1913 land misuse
f mésusage des terres
e utilización errónea de tierras
إساءة استغلال الأراضي

1914 land pollutant
f polluant du sol
e contaminante del suelo
ملوث التربة

1915 land pollution
f pollution du sol
e contaminación de tierra
تلوث الارض

1916 land pressure
f demande pressante de terres
e demanda apremiante de tierras
ضغط الطلب على الأرض

1917 land race
f cultivar traditionnel
e variedad natural
النباتات الأصلية

1918 land race conservation
f conservation des espèces primitives
e conservación de variedades naturales
حفظ الأنواع الأصلية

1919 land reclamation
f restauration des terres
e regeneración de tierras
استصلاح الأراضي

1920 land rehabilitation
f remise en état des terres
e regeneración de tierras
إصلاح الأرض

1921 land restoration
f restauration des sols
e regeneración de tierras
استعادة خصوبة الأرض

1922 land subsidence
f affaissement du sol
e subsidencia del suelo
هبوط الأرض

1923 land surface temperature
f température superficielle des terres
e temperatura de la superficie terrestre
درجة حرارة سطح الأرض

1924 land tenure (system)
f régime foncier
e sistema de tenencia de tierras
نظام حيازة الأراضي

1925 land transportation
f transports terrestres
e transporte terrestre
النقل البرى

1926 land use
f emploi des terres
e uso de la tierra
استخدام الأراضي

1927 land use classification
f classification de l'usage du sol
e clasificación del uso de las tierras
تصنيف استخدام الارض

1928 land use planning
f aménagement du territoire
e planificación del uso del suelo
تخطيط استخدام الأرض

1929 land values
f valeurs de la terre
e valor de las tierras
قيم الأرض

1930 land-based resources
f ressources de la terre
e recursos situados en tierra
موارد أرضية

1931 landfill gas
f gas d'enfouissement
e gas de vertedero
غاز مدافن القمامة

1932 landfilling
f mise en décharge; remblayage
e eliminación de desechos en vertederos
دفن القمامة

1933 landlessness
f privation des terres
e carencia de tierra
عدم ملكية الأراضي

1934 land-ocean interactions
f interactions continent-océan
e interacciones tierra-océano
التفاعلات بين البر والبحر

1935 landscape diversity
f diversité des sites
e diversidad de paisajes
تنوع المناظر الطبيعية

1936 landscape ecology
f écologie des paysages
e ecología de los paisajes
ايكولوجيا المناظر الطبيعية

1937 landscape protected area
f site protégé
e paisaje protegido
محمية مناظر طبيعية

1938 landscape renovation
f remodelage du paysage
e remodelación del paisaje
تجديد المناظر الطبيعية

1939 landscape restoration
f restauration des sites
e restauración del paisaje
استعادة جمال المناظر الطبيعية

1940 landscaped area
f espace vert aménagé
e espacio abierto acondicionado
منطقة مجملة

1941 landslide
f glissement de terrain
e deslizamiento de tierra
انهيال ارضي

1942 land-use management
f aménagement du territoire
e ordenación de las tierras
إدارة استخدام الأراضي

1943 lapse rate
f gradient thermique vertical
e gradiente térmico vertical
معدل التفاوت في درجة الحرارة

1944 large stem cuttings
f plançons
e estaquillas de tallo
قطع الجذوع الكبيرة

1945 large-pollution
f pollution diffuse
e contaminación difusa
تلوث واسع الانتشار

1946 larval
f zoé
e larval
متعلق بطور اليرقة

1947 laser station
f station laser
e estación de láser
محطة ليزر

1948 lateral transport
f transport latéral
e transporte lateral
نقل جانبي

1949 latitudinal zone
f zone de latitude
e zona de latitud
منطقة خط عرض

1950 laughing gas
f gaz hilarant
e gas hilarante
الغاز المضحك

1951 laundering
f blanchisserie
e lavanderías
غسل الثياب

1952 law-making
f activités créatrices de droit
e proceso legislativo
التشريع

1953 leach field
f champ d'épandage
e campo de aplicación (de fangos cloacales)
حقل تنقية مياه المجاري

1954 leachate
f lixivat
e lixiviado
السائل المرشح

1955 leaching
f lixiviation
e lixiviación
نضح

1956 lead contamination
f contamination par le plomb
e contaminación por plomo
تلوث بالرصاص

1957 lead poisoning
f intoxication par le plomb
e intoxicación por plomo
التسمم بالرصاص

1958 leaded
f au plomb
e con plomo
يحتوى على رصاص

1959 leader
f pousse
e guía
قمة الشجرة

1960 lead-free
f sans plomb
e sin plomo
خال من الرصاص

1961 leaf drop
f chute des feuilles
e caída de las hojas
سقوط الأوراق

1962 leaf necrosis
f nécrose foliaire
e necrosis foliar
موت الأوراق

1963 leaf season
f foliation
e foliación
موسم الإيراق

1964 leaf surface
f surface foliaire
e superficie foliar
سطح الورقة

1965 leaf-area index
f indice foliaire
e índice de superficie foliar
دليل كثافة الغطاء النباتي

1966 leak (age) (flow) rate
f débit de fuite
e caudal de pérdida
معدل التسرب

1967 lean-burn engine
f moteur à mélange pauvre
e motor de mezcla pobre
محرك قليل الاستهلاك لوقود الاحتراق

1968 leather industry
f industrie du cuir
e industria del cuero
صناعة الجلد

1969 lethal
f létal
e letal
مميت

1970 lethal concentration
f concentration létale
e concentración letal
تركيز مميت

1971 lethal dose
f dose létale
e dosis letal
جرعة مميتة

1972 level of compliance
f degré d'observation
e grado de cumplimiento
مستوى الامتثال

1973 levy on environmentally harmful consumption
f taxe sur la consommation préjudiciable à l'environnement
e gravamen del consumo perjudicial para el medio ambiente
ضريبة الاستهلاك الضار بيئيا

1974 liability
f responsabilité
e responsabilidad
المسؤولية

1975 liability for environmental damage
f responsabilité en matière de préjudice écologique
e responsabilidad por daños al medio ambiente
المسؤولية عن الضرر البيئى

1976 liability for marine accidents
f responsabilité des accidents maritimes
e responsabilidad por accidentes marítimos
مسؤولية عن الحوادث البحرية

1977 liability for nuclear damages
f responsabilités en matière de dommages nucléaires
e responsabilidad por daños nucleares
مسؤولية عن الأضرار النووية

1978 liability system
f système de responsabilité
e régimen de responsabilidad civil
نظام المسؤولية

1979 licence to discharge
f autorisation de rejet
e permiso de descarga
إذن بتصريف

1980 licensed process
f procédé exploité sous licence
e proceso obtenido por licencia
عملية مرخصة

1981 life-support capacity
f capacités biologiques
e capacidad de sustentar la vida
قدرة على دعم الحياة

1982 life cycle
f cycle d'évolution
e ciclo vital
دورة حياة

1983 life support(ing) system
f système d'entretien de la vie
e sistema sustentador de la vida
جهاز المحافظة على الحياة

1984 lifestyles
f modes de vie
e estilos de vida
أساليب الحياة

1985 light-duty vehicle
f véhicule utilitaire léger
e vehículo liviano
مركبة حمولات خفيفة

1986 light ends
f coupes légères
e fracciones ligeras
المتطايرات

1987 light radiation
f rayonnement lumineux
e radiación luminosa
أشعة ضوئية

1988 light stabilizing (agent)
f photostabilisant
e fotoestabilizador
عامل التثبيت الضوئى

1989 lighting
f éclairage
e alumbrado
برق

1990 lightly pigmented skin
f peau claire
e piel clara
بشرة فاتحة اللون

1991 limb scanning
f balayage du disque terrestre
e barrido del disco terrestre
مسح اشعاعى للأرض

1992 lime potential
f potentiel calcaire
e potencial en cal
الحمل القلوى

1993 lime treatment
f traitement à la chaux
e tratamiento con cal
معالجة بالجير

1994 limestone land
f terrains calcaires
e terreno calcáreo
أرض جيرية

1995 limestone scrubbing process
f procédé de lavage à la chaux
e lavado con cal
عملية غسل بالحجر الجيرى

1996 limestone soil
f sol calcaire
e suelo calcáreo
تربة جيرية

1997 limnetic zone
f zone limnétique
e zona limnética
منطقة المياه العذبة العميقة

1998 limnology
f limnologie
e limnología
علم المياه العذبة

1999 line source
f source linéaire
e fuente lineal
مصدر خطى

2000 lined
f étanchéifié
e revestido
مبطن

2001 liquefied gas
f gaz liquéfiés
e gases licuados
غاز مسال

2002 liquefied petroleum gas
f gaz de pétrole liquéfié
e gas de petróleo licuado
غاز نفطى مسيل

2003 liquid ash extraction
f extraction des cendres par voie humide
e extracción de cenizas por vía húmeda
استخلاص الرماد السائل

2004 liquid chromatography
f chromatographie en phase liquide
e cromatografía en fase líquida
الفصل اللونى بالسوائل

2005 liquid dye laser
f laser à colorant
e láser de colorante líquido
ليزر ملون

2006 liquid wastes
f déchets liquides
e desechos líquidos
نفايات سائلة

2007 lithosphere
f lithosphère
e litosfera
القشرة الأرضية

2008 litter
f couverture morte
e cubierta muerta
فرش حرجى

2009 litter
f petits déchets
e basura
قمامة

2010 litter collection
f ramassage des petits déchets
e recolección de basura
جمع القمامة

2011 littering
f abandon de détritus
e tirar basura
إلقاء القمامة

2012 littoral environment
f milieu litttoral
e medio litoral
بيئة ساحلية

2013 live species
f espèces vivantes
e especie (s) viva (s)
انواع حية

2014 livelihood
f moyens de subsistance
e medios de subsistencia
أسباب المعيشة

2015 livestock buildings
f bâtiments d'élevage
e construcciones para ganado
حظائر الماشية

2016 livestock system
f système d'élevage
e sistema de ganadería
نظام تربية المواشى

2017 living environment
f cadre de vie
e medio vital
البيئة المعيشية

2018 living marine resources
f ressources biologiques maritimes
e recursos marinos vivos
موارد بحرية حية

2019 living organism
f être vivant
e organismo vivo
كائن حى

2020 living resources
f ressources biologiques
e recursos vivos
الموارد الحية

2021 load
f charge
e carga
كمية

2022 load per unit of mass
f charge massique
e carga por unidad de masa
الحمل فى وحدة الكتلة

2023 loading
f teneur concentration
e cantidad
كمية التركيز

2024 local
f local
e local
محلى

2025 local (emission) source
f source (d'émissions polluantes) locale
e fuente local (de contaminación)
مصدر محلى لإنبعاث الملوثات

2026 local air pollution
f pollution atmosphérique à l'échelon local
e contaminación atmosférica local
تلوث هوائى محلى

2027 **local environment**
f cadre de vie
e medio ambiente local
بيئة محلية

2028 **local indigenous crops**
f plantes cultivées localement
e cultivos autóctonos
محاصيل محلية أصلية

2029 **local materials for building**
f matériaux de contruction locaux
e materiales de construcción
مواد محلية للبناء

2030 **location-specific research**
f recherche strictement localisée
e investigaciones de enfoque local
بحث على موقع معين

2031 **locust control**
f lutte antiacridienne
e lucha contra la langosta
مكافحة الجراد

2032 **locust outbreak**
f infestation acridienne
e invasión de langostas
تفشى الجراد

2033 **locust plague**
f invasion de criquets pélerins
e plaga de la langosta
غزو الجراد

2034 **locust swarm**
f nuée de criquets pélerins
e nube de langostas
سرب الجراد

2035 **lofting plume**
f panache tourmenté vers le haut
e penacho elevado
عمود دخان مرتفع

2036 **logging**
f abattage
e corta
قطع الأشجار

2037 **long-lived toxic substance**
f substance toxique persistante
e sustancia tóxica de larga vida
مادة سامة معمرة

2038 **long-lived waste**
f déchets à vie longue
e desechos de período prolongado
نفايات معمرة

2039 **long-range transport(ation)**
f transport à longue distance
e transporte a larga distancia
نقل بعيد المدى

2040 **long-term effects of pollutants**
f effets à long terme des polluants
e efectos a largo plazo de contaminantes
آثار طويلة الأجل للملوثات

2041 **long-term exposure**
f exposition de longue durée
e exposición prolongada
تعرض لفترات طويلة

2042 long-term forecasting
f prévisions à long terme
e predicciones a largo plazo
توقع طويل الأجل

2043 long-term trends
f tendances à long terme
e tendencias a largo plazo
اتجاهات طويلة الأجل

2044 long-wave (length) radiation
f rayonnement de grande longueur d'onde
e radiación de onda larga
إشعاع طويل الموجة

2045 looping
f panache en boucles
e penacho serpenteante
الحركة الأنشوطية للدخان

2046 loss of containment
f défaillance du confinement
e falla del confinamiento
فقد الاحتواء

2047 loss of genetic variations
f perte de variantes génétiques
e pérdida de variantes genéticas
نقص التنوع الوراثى

2048 loss of a cropping season
f perte d'une récolte
e pérdida de una cosecha
خسارة موسم الزراعة

2049 loss of biodiversity
f diminution de la diversité biologique
e disminución de la diversidad biológica
نقص التنوع البيولوجى

2050 loss of coastal land
f disparition de terres côtières
e pérdida de tierras costeras
فقد الأراضي الساحلية

2051 loss of species
f disparition d'espèces
e desaparición de especies
فقد الأنواع

2052 loss of vegetation
f dépérissement de la végétation
e pérdida de vegetación
فقد الغطاء الخضرى

2053 low alcohols
f alcools légers
e alcoholes inferiores
الكحوليات الخفيفة

2054 low latitudes
f basses latitudes
e latitudes bajas
خطوط العرض المنخفضة

2055 low pollution
f pollution réduite
e contaminación reducida
تلوث قليل

2056 low toxic concentration
f teneur toxique minimale
e concentración tóxica mínima
سموم قليلة التركيز

2057 low toxic dose
f dose toxique minimale
e dosis tóxica mínima
جرعة سامة منخفضة

2058 low water level
f étiage
e marea baja
أدنى منسوب الماء

2059 low(er) stratosphere
f basse stratosphère
e estratosfera inferior
الطبقة السفلى من الستراتوسفير

2060 low- and non-waste technologies
f techniques peu polluantes ou sans déchets
e tecnologías de desechos escasos o nulos
التكنولوجيات القليلة والعديمة النفايات

2061 low-cost housing
f logements à faible coûts
e viviendas económicas
إسكان منخفض التكاليف

2062 low-cost technology
f technique peu onéreuse
e técnica de bajo costo
تكنولوجيا منخفضة التكلفة

2063 low-emission
f à faible taux d`émission
e de bajo nivel de emisiones
قليل الانبعاث

2064 low-emission technology
f technique peu polluante
e tecnología poco contaminante
تكنولوجيا قليلة الانبعاث

2065 low-energy consumption economy
f économie consommant moins d`énergie
e economía de bajo consumo de energía
اقتصاد منخفض الاستهلاك للطاقة

2066 lower limit of detectability
f limite de détection
e cantidad detectable mínima
الحد الأدنى للكشف

2067 lower polar stratosphere
f basse stratosphère polaire
e estratosfera polar inferior
الستراتوسفير القطبية السفلية

2068 lowland area
f zone de faible altitude
e zona de tierras bajas
منطقة أراض منخفضة

2069 lowlands
f plaines;basses terres
e tierras bajas
أراض منخفضة

2070 low lethal concentration
f teneur létale minimale
e concentración letal mínima
أدنى تركيز مميت

2071 low-level waste
f déchets de faible activité
e desechos poco activos
نفايات منخفضة النشاط

2072 low-lying coastal area
f plaine côtière
e zona de litoral bajo
منطقة ساحلية منبسطة

2073 low-lying country
f pays de faible altitude
e país de baja altitud
بلد منخفض

2074 low-nutrient waters
f eaux pauvres en nutriments
e aguas de bajo contenido de nutrientes
مياه قليلة المغذيات

2075 low-waste technology
f techniques peu polluantes
e tecnología de desechos escasos
تكنولوجيا قليلة النفايات

M

2076 macroclimate
f macroclimat
e macroclima
المناخ الكلى

2077 magnetosphere
f magnétosphère
e magnetósfera
الغلاف المغناطيسى

2078 magnitude
f magnitude
e magnitud
شدة

2079 main water table
f surface de la nappe phréatique
e nivel fréatico principal
منسوب المياه الجوفية الرئيسى

2080 maintenance of environmental quality
f maintien de la qualité de l'environnement
e mantenimiento de la calidad del medio ambiente
المحافظ على نوعية البيئة

2081 major constituent
f constituant majoritaire
e constituyente principal
مكّون رئيسى

2082 malaria
f paludisme
e paludismo
ملاريا

2083 malignant melanoma
f mélanome malin
e melanoma maligno
ورم خبيث

2084 malnutrition
f malnutrition
e malnutrición
سوء التغذية

2085 malpractice
f faute professionnelle
e negligencia profesional
ممارسة مهنية سيئة

2086 mammals
f mammifères
e mamíferos
ثدييات

2087 man-induced degradation
f dégradation causée par l'homme
e degradación provocada por el hombre
تدهور بفعل الإنسان

2088 managed plantation
f plantation aménagée
e plantación
مزرعة مدارة

2089 mangrove
f mangrove
e manglar
أشجار المانجروف

2090 mangrove swamps
f mangroves marécageux
e manglares
مستنقعات المانجروف

2091 man-made disasters
f catastrophes causées par l'homme
e desastres causados por el hombre
كوارث من صنع الإنسان

2092 man-made dust
f poussière d'origine industrielle
e polvo generado por el hombre
غبار صناعى

2093 man-made environment
f environnement aménagé
e medio (ambiente) creado por el hombre
بيئة اصطناعية

2094 man-made fibers
f fibres chimiques
e fibras manufacturadas
ألياف صناعية

2095 man-made forest
f forêt artificielle
e bosque artificial
غابات مزروعة

2096 man-made hazard
f risque anthropique
e riesgo creado por el hombre
خطر من صنع الإنسان

2097 man-made landscape
f paysage urbain
e paisaje creado por el hombre
منظر طبيعى اصطناعي

2098 man-made pollutant
f polluant artificiel
e contaminante generado por el hombre
ملوث اصطناعي

2099 man-made pollution
f pollution industrielle
e contaminación de origen humano
تلوث ناجم عن الأنشطة البشرية

2100 manure
f fumier
e estiércol
السماد الطبيعى

2101 manure gas
f gas de fumier
e gas de estercolero
غاز السماد الطبيعى

2102 mapping
f cartographie
e cartografía
رسم الخرائط

2103 mapping of vegetation
f cartographie de la végétation
e cartogafía de la vegetación
خرائط النباتات

2104 marginal farmer
f paysan cultivant des terres à rendement marginal
e agricultor marginal
مزارع يستغل الأراضي الحديّة

2105 marginal land
f terre de faible rendement
e tierras marginales
أراضٍ حدية

2106 marginal productivity of capital
f taux marginal de productivité du capital
e productividad marginal del capital
الإنتاجية الحدية لرأس المال

2107 marginal treatment costs
f coûts marginaux d'épuration
e costos marginales de tratamiento
تكاليف المعالجة الحدية

2108 mariculture
f mariculture
e cultivo marino
تربية الأحياء البحرية

2109 marine
f marin
e marino
بحرى

2110 marine biodiversity
f diversité biologique des mers
e diversidad biológica marina
التنوع البيولوجى البحرى

2111 marine biology
f biologie marine
e biología marina
علم الأحياء البحرية

2112 marine biota
f biote marin
e biota marina
الأحياء البحرية

2113 marine body
f étendue marine
e masa marina
منطقة بحرية

2114 marine chemistry
f chimie de la mer
e química del mar
الكيمياء البحرية

2115 marine climate
f climat maritime
e clima marítimo
مناخ بحرى

2116 marine climatology
f climatologie maritime
e climatología marina
علم المناخ البحرى

2117 marine conservation areas
f zones de conservation marine
e conservación de áreas marinas
مناطق الصيانة البحرية

2118 marine disposal
f rejet en mer
e eliminación en el mar
التخلص فى البحر

2119 marine ecosystems
f milieux marins
e ambientes marinos
نظم إيكولوجية بحرية

2120 marine engineering
f génie marin
e ingeniería marítima
هندسة بحرية

2121 marine environment
f environnement marin
e medio marino
بيئة بحرية

2122 marine fishery
f pêche en mer
e pesca marítima
صيد الأسماك البحرية

2123 marine food web
f réseau trophique marin
e red alimentaria marina
شبكة غذائية بحرية

2124 marine fuels
f carburants des navires
e combustibles navales
وقود المركبات البحرية

2125 marine geology
f géologie marine
e geología marina
جيولوجيا بحرية

1226 marine incineration
f incinération en mer
e incineración en el mar
الحرق فى البحر

2127 marine life
f faune et flore marines
e fauna y flora marinas
الحياة البحرية

2128 marine meteorological services
f assistance météorologique aux activités maritimes
e servicios meteorológicos marinos
خدمات الأرصاد الجوية البحرية

2129 marine monitoring
f surveillance du milieu marin
e vigilancia marina
رصد بحرى

2130 marine pollution
f pollution marine
e contaminación de los mares
تلوث بحرى

2131 marine pollution from land-based sources
f pollution marine d'origine tellurique
e contaminación marina procedente de fuentes terrestres
التلوث البحرى من مصادر برية

2132 marine resources conservation
f conservation des ressources marines
e conservación de recursos
صيانة الموارد البحرية

2133 marine sciences
f sciences de la mer
e ciencias marinas
علوم البحار

2134 marine sediments
f sédiments marins
e sedimentos marinos
رواسب بحرية

2135 maritime transport
f transport maritime
e transporte marítimo
نقل بحرى

2136 market-based incentive
f incitation commerciale
e incentivo comercial
حافز سوقى

2137 marketing with green labelling
f commercialisation avec label vert
e comercialización con etiqueta verde ecológica
تسويق بوضع علامات بيئية

2138 marshland
f marais
e marisma
مستنقعات

2139 mass balance
f bilan massique
e balance de masas
توازن كتلى

2140 mass mixing ratio
f rapport de mélange
e relación de mezcla
نسبة الخلط

2141 mass spectrometer
f spectromètre de masse
e espectrómetro de masas
مطياف كتلى

2142 mass transfer coefficient
f coefficient de transfert de matière
e coeficiente de transferencia de masas
معامل انتقال الكتلة

2143 materials science
f sciences des matériaux
e ciencia de los materiales
علم المواد

2144 matrix effects
f effets de matrice
e efectos matriciales
آثار المصفوفة

2145 matrix of energy supply
f filière énergétique
e conjunto de fuentes energéticas
مصفوفة إمدادات الطاقة

2146 mature forest stand
f peuplement forestier adulte
e rodal maduro
غابة ناضجة

2147 maximum allowable concentration
f concentration maximale admissible
e concentración máxima admisible
أقصى ارتفاع مسموح به

2148 maximum allowable dose
f dose maximale admissible
e dosis máxima admisible
أقصى جرعة مسموح بها

2149 maximum desirable concentration
f teneur maximale souhaitable
e concentración máxima deseable
أقصى تركيز مسموح به

2150 maximum emission concentration
f teneur maximale à l'émission
e concentración máxima
التركيز الأقصى للانبعاثات

2151 maximum sustainable yield
f rendement équilibré maximal
e rendimiento máximo permisible
الغلة القصوى المستدامة

2152 maximum tolerable concentration
f teneur maximale admissible
e concentración máxima tolerable
أقصى تركيز يمكن تحمله

2153 maximum tolerated dose
f dose maximale tolérée
e dosis máxima tolerada
أقصى جرعة يمكن تحملها

2154 mean pressure chart
f carte barométrique moyenne
e carta de presiones medias
جدول متوسط الضغط الجوى

2155 mean sea level
f niveau moyen de la mer
e nivel medio del mar
متوسط مستوى سطح البحر

2156 mean surface temperature
f température moyenne en surface
e temperatura superficial media
متوسط درجة الحرارة عند السطح

2157 meander belt
f lit des méandres
e zona cubierta por los meandros
منطقة التعرج النهرى

2158 measurement of air pollution from space
f mesure de la pollution atmosphérique à partir de l'espace
e medición de la contaminación atmosférica desde el espacio
قياس التلوث الجوى من الفضاء

2159 mechanical gathering
f ramassage par voie mécanique
e recolección mecánica
جمع آلى

2160 mechanical engineering
f génie mécanique
e ingeniería mecánica
هندسة ميكانيكية

2161 median lethal dose
f dose létale moyenne
e dosis mediana letal
نصف الجرعة المميتة

2162 medical entomologist
f patho-entomologiste
e entomólogo médico
أخصائي فى علم الحشرات الطبى

2163 medical treatment
f traitements médicaux
e tratamiento médico y cuidado de la salud
علاج طبى

2164 medicinal plants
f plantes médicinales
e plantas medicinales
نباتات طبية

2165 medium reactivity hydrocarbons
f hydrocarbures peu dégradables
e hidrocarburos de reactividad media
مواد هيدروكربونية متوسطة النشاط

2166 medium-level waste
f déchets de moyenne activité
e desechos de actividad media
نفايات متوسطة النشاط

2167 medium-risk GMOs
f OGM à risque moyen
e organismos modificados genéticamente de riesgo medio
كائنات محورة وراثيا متوسطة الخطر

2168 megacity
f mégalopole
e megalópolis
مدينة ضخمة

2169 megathermal period
f période mégathermique
e periodo megatérmico
فترة الحرارة العالية

2170 melanoma
f mélanome
e melanoma
ورم قتامى

2171 melanoma skin cancer
f cancer de la peau avec présence de mélanome(s)
e cáncer de piel con melanoma
سرطان جلدى قتامى

2172 membrane filter
f membrane filtrante
e filtro (de) membrana
مرشح غشائى

2173 membrane technology
f technologie des membranes
e tecnología de membranas
تكنولوجيا الأغشية

2174 memory effect
f effet mémoire
e efecto memoria
اثر على الذاكرة

2175 mercury contamination
f contamination par le mercure
e contaminación por mercurio
تلوث بالزئبق

2176 meridional ozone transport
f transport méridien de l'ozone
e transporte meridiano de ozono
الانتقال الطولى للأوزون

2177 meridional transfer
f transfert méridien
e transferencia meridiana
انتقال طولى

2178 mesoclimate
f mésoclimat
e mesoclima
مناخ متوسط النطاق

2179 mesotrophic lake
f lac mésotrophe
e lago mesotrófico
بحيرة ذات مغذيات متوسطة

2180 metabolic needs
f besoins essentiels de l'organisme
e necesidades metabólicas
احتياجات التفاعل الحيوى

2181 metabolism
f métabolisme
e metabolismo
تفاعل حيوى

2182 metabolism of pesticides
f métabolisme des pesticides
e metabolismo de plaguicidas
التفاعل الحيوى لمبيدات الأعشاب

2183 metal degreasing
f dégraissage des pièces métalliques
e desengrasado de metales
إزالة الشحوم من الفلزات

2184 metal drum
f fût métallique
e tonel metálico
برميل معدنى

2185 metal finishing
f finition des métaux
e finición de metales
تهذيب المعادن

2186 metal hydroxide
f hydroxyde métallique
e hidróxido metálico
هيدروكسيد فلزى

2187 metal ion
f ion métallique
e ion metálico
أيون فلزى

2188 metal oxides
f oxydes métalliques
e óxidos metálicos
أكاسيد فلزية

2189 metal plating
f métallisation
e metalización
تغطية بألواح معدنية

2190 metal salts
f sels métalliques
e sales metálicas
أملاح فلزية

2191 metal smelting
f fusion des métaux
e fundición de metales
صهر المعادن

2192 meteorological radar
f radar météorologique
e radar meteorológico
رادار الأرصاد الجوية

2193 meteorological rocket
f fusée météorologique
e cohete meteorológico
صاروخ الرصد الجوى

2194 meteorological extremes
f conditions météorologiques extrêmes
e extremos meteorológicos
الحدود القصوى للأحوال الجوية

2195 meteorological satellite
f satellite météorologique
e satélite meteorológico
ساتل للأرصاد الجوية

2196 meteorology
f météorologie
e meteorología
علم الأرصاد الجوية

2197 metering of the air-fuel ratio
f dosage du mélange air-carburant
e calibración de la mezcla aire-combustible
قياس نسبة الهواء الى الوقود

2198 methane fermentation
f fermentation méthanogène
e fermentación metánica
تخمير الميثان

2199 methane recovery
f récupération du méthane
e recuperación de metano
استعادة الميثان

2200 methane sink
f puits de méthane
e sumidero del metano
بالوعة ميثان

2201 method of disposal
f mode d'élimination
e método de eliminación
طريقة تخلص

2202 methylene blue test
f épreuve au bleu de méthylène
e prueba de azul de metileno
كشف بالميثلين الأزرق

2203 microaerophilic condition
f condition de microaérophilie
e condición microaerófila
ظرف هوائى مجهرى

2204 microbial leaching
f lessivage par les bactéries
e lixiviación microbiana
نض جرثومى

2205 microbial polymer
f polymère microbien
e polímero microbiano
بوليمر جرثومى

2206 microbial resources
f ressources microbiennes
e recursos microbianos
موارد جرثومية

2207 microbial strain
f souche microbienne
e capa microbiana
سلالة جرثومية

2208 microbial system
f système microbien
e sistema microbiano
نظام جرثومى

2209 microbiology
f microbiologie
e microbiología
علم الأحياء الدقيقة

2210 microclimate
f microclimat
e microclima
مناخ منطقة صغيرة

2211 microclimate effects
f effets microclimatiques
e efectos de microclimas
آثار مناخ موضعي

2212 microclimatology
f microclimatologie
e microclimatología
علم المناخ الموضعى

2213 microcosm test
f essai en microécosystème
e ensayo en microcosmos
اختبار تطور الكائنات الصغيرة

2214 microfauna
f microfaune
e microfauna
الحيوانات الصغيرة جداً

2215 microflora
f microflore
e microflora
النباتات المجهرية

2216 micronutrient
f micronutriment
e microelemento nutritivo
مغذيات دقيقة

2217 micronutrient malnutrition
f carences en micronutriments
e insuficiencia de micronutrientes
نقص المغذيات الدقيقة

2218 micro-organism
f micro-organisme
e microoganismo
كائنات مجهرية

2219 micropollutant
f micropolluant
e microcontaminante
ملوثات دقيقة

2220 microwave measurements
f mesures en hyperfréquences
e mediciones por microondas
قياسات بموجات متناهية القصر

2221 microwave radio meter
f radiomètre à hyperfréquences
e radiómetro de microondas
مقياس إشعاع متناهى القصر

2222 microwave window
f fenêtre des hyperfréquences
e ventana de microondas
نافذة الموجات المتناهية القصر

2223 middle atmosphere
f atmosphère moyenne
e atmósfera media
طبقات الغلاف الجوى الوسطى

2224 midnight dumping
f rejet sauvage
e vertimiento ilegal
إلقاء النفايات غير المشروع

2225 migrant workers
f travailleurs migrants
e trabajadores emigrantes
عمال مهاجرون

2226 migratory paths
f voies de migration
e rutas de migración
مسارات الهجرة

2227 migratory settlement
f habitat précaire
e concentración provisional
استيطان عابر

2228 migratory species
f espèce(s) migratrice(s)
e especie migratoria
الأنواع المهاجرة

2229 migratory locust
f criquet migrateur
e langosta migratoria
الجراد المهاجر

2230 military activity
f activités militaires
e actividad militar
نشاط عسكرى

2231 mine filling
f remblayage de mines
e rellenado de minas
ردم المناجم

2232 mine tailings
f déchets de mine
e desechos de minería
مخلفات المناجم

2233 mineral industry
f industrie minéralogique
e industria mineral
صناعة المعادن

2234 mineral resources
f ressources minérales
e recursos minerales
موارد معدنية

2235 minihole
f trou d'ozone secondaire
e agujero secundario en la capa de ozono
ثقب صغير فى الأزون

2236 mini-hydro system
f système mini -hydraulique
e sistema hidroeléctrico de pequeña escala
نظام مائى صغير

2237 minimum effective dose
f dose minimale efficace
e dosis eficaz mínima
اقل جرعة مؤثرة

2238 minimum surprise approach
f méthode de la surprise minimum
e método de sorpresas mínimas
نهج تفادى الخطر

2239 mining
f industrie minière
e minería
تعدين

2240 mining engineering
f génie minier
e ingeniería minera
هندسة التعدين

2241 mining geology
f géologie minière
e geología minera
جيولوجيا التعدين

2242 mining wastes
f déchets des industries minières
e desechos de minería
نفايات التعدين

2243 minor constituent of sea water
f constituant mineur de l'eau de mer
e constituyentes menores del agua de mar
مكون بسيط لماء البحر

2244 minor constituent of the atmosphere
f constituant minoritaire de l'atmosphère
e constituyentes menores de la atmósfera
مكون بسيط للغلاف الجوى

2245 minor crops
f menues cultures
e cultivos menores
محاصيل ثانوية

2246 minor forest product
f menu produit forestier
e producto forestal secundario
منتج حرجى ثانوى

2247 minorities
f minorités
e minorías
أقليات

2248 mishap
f accident
e accidente
حادث

2249 mission-orientated research
f recherche thématique
e investigación aplicada
بحث موضوعى

2250 mis-use
f mésusage
e mala utilización
إساءة استعمال

2251 mitigate
f atténuer
e mitigar
خفف

2252 mitigation measures
f mesures d'atténuation des effets
e medidas de mitigación
تدابير التخفيف

2253 mitigation of climate change
f atténuation du changement climatique
e mitigación de los cambios climáticos
الحد من تغير المناخ

2254 mitigation of natural disasters
f atténuation des catastrophes naturelles
e mitigación de los desastres naturales
الحد من الكوارث الطبيعية

2255 mitigation procedure
f procédure d'atténuation
e procedimiento de mitigación
طريقة التخفيف

2256 mixed fuels
f mélange de combustibles
e combustibles diversos
خليط وقودى

2257 mixing bowl pollutant
f polluant réactif
e contaminante reactivo
ملوثات متفاعلة

2258 mixing ratio by volume
f rapport de mélange en unités de volume
e relación de mezcla volumétrica
نسبة الخلط بالحجم

2259 mobile homes
f habitations mobiles
e viviendas móviles
منازل نقالى

2260 mobile source (of emissions)
f source mobile (d'émissions)
e fuente móvil (de emisiones)
مصدر (انبعاثات) متحرك

2261 mobilization
f mobilisation
e movilización
تيسير القابلية للحركة

2262 mode of toxicity
f mécanisme d'action toxique
e mecanismo de acción tóxica
طريقة التسمم

2263 model calibration
f calibration modèle
e calibración de modelos
معايرة النموذج

2264 model ecosystem
f écosystème expérimental
e ecosistema modelo
نظام ايكولوجى نموذجى

2265 model output diagnosis
f sortie directe de modèle
e análisis de los resultados del modelo
تشخيص نواتج النموذج

2266 model output statistics
f statistique de sortie de modèle
e estadísticas de los resultados del modelo
إحصاءات نواتج النموذج

2267 model predictions
f prévisions du modèle
e predicción basada en modelos
التنبوءات المستمدة من النموذج

2268 model validation
f vérification du modèle
e validación del modelo
التحقق من سلامة النموذج

2269 modeller
f modéliseur
e constructor de modelos
صانع نماذج

2270 modelling
f modélisation
e modelos
وضع النماذج

2271 modelling of climate change
f modélisation de l'évolution du climat
e modelado de los cambios climáticos
صنع نموذج لتغير المناخ

2272 moder
f moder
e moder
دبال بنى ناعم

2273 moderate utilization of natural resources
f exploitation judicieuse des ressources naturelles
e aprovechamiento prudente de los recursos naturales
الاستغلال المعتدل للموارد الطبيعية

2274 modern liquid chromatography
f chromatographie en phase liquide moderne
e cromatografía moderna en fase líquida
الفصل اللونى الحديث بالسوائل

2275 modified trait
f caractère modifié
e característica modificada
سمة محورة

2276 moist meadow
f prairie humide
e prado húmedo
مرج رطب

2277 moisture content
f état hygrométrique
e contenido de humedad
نسبة الرطوبة

2278 moisture content profile
f profil hydrique
e perfil hídrico
مقطع رأسي لنسبة الرطوبة

2279 moisture proof
f résistant à l'humidité
e resistente a la humedad
مقاوم للرطوبة

2280 mole fraction
f fraction molaire
e fracción molar
كسر جزيئى

2281 molecular abundance
f fraction moléculaire
e fracción molecular
النسبة الجزيئية

2282 molecular biology
f biologie moléculaire
e biología molecular
بيولوجيا الجزيئات

2283 molecular chlorine
f chlore moléculaire
e cloro molecular
كلور جزيئى

2284 molecular scattering
f diffusion moléculaire
e dispersión molecular
تشتت جزيئى

2285 molecular technologies
f technologies moléculaires
e tecnologías moleculares
تكنولوجيات الجزيئات

2286 molecule breakdown
f fragmentation d'une molécule
e fragmentación molecular
تكسر الجزيء

2287 monitor
f détecteur
e detector
جهاز رصد

2288 monitor the weather
f suivre l'évolution du temps
e observar el tiempo
رصد الحالة الجوية

2289 monitoring
f surveillance
e vigilancia
رصد

2290 monitoring criteria
f critères de surveillance
e criterios de vigilancia
معايير الرصد

2291 monitoring data
f données sur la surveillance
e elementos de vigilancia
بيانات الرصد

2292 monitoring equipment
f équipements de surveillance
e equipo de vigilancia
معدات رصد

2293 monitoring station
f station de surveillance
e estación de observación
محطة رصد

2294 monitoring techniques
f techniques et équipements de surveillance
e técnicas y equipos de vigilancia
تقنيات الرصد

2295 monolithic catalyst
f catalyseur monolithique
e catalizador monolítico
حفاز متجانس

2296 monsoon climatology
f climatologie des moussons
e climatología de los monzones
علم مناخ الرياح الموسمية

2297 monsoon dynamics
f dynamique des moussons
e dinámica de los monzones
ديناميات الرياح الموسمية

2298 monsoon-affected country
f pays de mousson (s)
e país afectado por los monzones
بلد متأثر بالرياح الموسمية

2299 monsoonal failure
f insuffisance de la mousson
e insuficiencia de los monzones
ضعف الرياح الموسمية

2300 mor
f humus brut
e mor;humus de pinaza
تربة دبالية حمضية

2301 most disaster-prone area
f région la plus exposée à des catastrophes
e zona más expuesta a desastres
أكثر المناطق تعرضا للكوارث

2302 motions of the atmosphere
f circulation atmosphérique
e circulación atmosférica
دوران الغلاف الجوى

2303 motor gasoline
f essence (pour) automobile
e gasolina para motores
بنزين المحركات

2304 motor vehicle emissions
f emissions des véhicules à moteur
e emisiones de vehículos automotores
انبعاثات المركبات

2305 motor vehicles
f automobiles
e vehículos automotores
مركبات

2306 motorcycles
f motocyclettes
e motocicletas
دراجات بخارية

2307 mould-release agent
f produit démoulant
e agente de desmoldeo
عامل إطلاق العفن

2308 moulting ground
f zone de mue
e terrenos de muda
منطقة طرح الشعر

2309 mountain ecosystem
f écosystèmes des montagnes
e ecosistemas de montañas
النظم الايكولوجية للجبال

2310 mountain glacier
f glacier
e glaciar (de montaña)
جليدية جبلية

2311 mountaineering
f alpinisme
e montañismo y alpinismo
تسلق الجبال

2312 moving curtain filter
f filtre à déroulement continu
e filtro de cortina móvil
مرشح ذو ستار متحرك

2313 muck soils
f humus
e tierra turbosa
تربة دبالية

2314 mudflow
f coulée de boue
e corriente de lodo
تدفق الوحل

2315 mudslide
f coulée de boues
e alud de lodo
انهيال وحلى

2316 mulch
f paillis
e cubierta orgánica
غطاء عضوى واق

2317 mull
f humus doux
e humus
دبال لين

2318 multicyclone device
f multicyclone
e ciclón múltiple
جهاز متعدد الدوامات

2319 multi-mode flow
f écoulement polyphasique
e flujo polifísico
دفق متعدد الطرق

2320 multiple scattering event
f cas de diffusion multiple
e suceso de dispersión múltiple
حالة تشتيت متعدد

2321 multiple sources
f sources multiples
e fuentes múltiples
مصادر متعددة

2322 multiplier effect
f effet multiplicateur
e efecto multiplicador
تأثير مضاعف

2323 multi-purpose tree
f arbre à usages multiples
e árbol de usos múltiples
شجرة متعددة الأغراض

2324 municipal sewage
f eaux d'égouts urbains
e aguas residuales urbanas
مياه المجاري الحضرية

2325 municipal engineering
f génie urbain
e obras públicas
هندسة البلديات

2326 municipal refuse disposal area
f zone d'élimination des ordures ménagères
e vertedero municipal
منطقة تخلص من النفايات الحضرية

2327 municipal solid waste
f déchets urbains solides
e residuos sólidos urbanos
النفايات الصلبة الحضرية

2328 municipal waste
f déchets municipaux
e desechos municipales
نفايات حضرية

2329 municipal waste indicator
f indicateur de déchets municipaux
e indicador de residuos urbanos
مؤشر نفايات البلديات

2330 municipal water distribution systems
f systèmes municipaux de distribution de l'eau
e acueducto municipal
شبكات توزيع المياه الحضرية

2331 mutagenesis
f mutagénèse
e mutagénesis
تولد الطفرات

2332 mutagens
f mutagènes
e mutágenos
مولدات الطفرات

2333 mutants
f mutants
e mutantes
طفرات

2334 mutated microorganisms release
f introduction accidentelle de micro-organismes modifiés(dans l'environnement)
e introducción de microorganismos modificados en el medio ambiente
إطلاق كائنات حية دقيقة مشوهة

N

2335 narrow-vision development
f développement à courte vue
e desarrollo con miras estrechas
تنمية تتسم بقصر النظر

2336 national conservation programmes
f programme de conservation nationale
e programas nacionales de conservación
برامج الصيانة الوطنية

2337 national legislation
f législations nationales
e legislación nacional
تشريع وطنى

2338 national park
f parc national
e parque nacional
روضة وطنية

2339 national reserves
f réserves nationales
e reservas nacionales
محتجزات وطنية

2340 native forest
f forêt naturelle
e bosque autóctono
غابة طبيعية

2341 native meadow
f prairie naturelle
e prado natural
مرج طبيعى

2342 natural assets
f actifs naturels
e activos naturales
أصول طبيعية

2343 natural capital
f capital naturel
e capital natural
رأس مال طبيعى

2344 natural cover
f couvert naturel
e cubierta natural
غطاء طبيعى

2345 natural degradation
f dégradation par action des agents naturels
e degradación natural
تدهور طبيعى

2346 natural disaster reduction
f prévention des catastrophes naturelles
e reducción de los efectos de los desastres
الحد من الكوارث الطبيعية

2347 natural drainage systems
f systèmes naturels de drainage
e sistemas naturales de drenaje
نظم التصريف الطبيعية

2348 natural environment
f milieu naturel
e medio natural
بيئة طبيعية

2349 natural environment rehabilitation
f restauration des milieux naturels
e restauración del medio natural
إصلاح البيئة الطبيعية

2350 natural environmental stress
f agression naturelle
e tensión ambiental natural
إجهاد البيئة الطبيعية

2351 natural fertilizers
f engrais naturels
e abonos naturales
أسمدة عضوية

2352 natural fibers
f fibres naturelles
e fibras textiles naturales
ألياف طبيعية

2353 natural gas exploration
f exploration du gaz naturel
e exploración de gas natural
استكشاف الغاز الطبيعى

2354 natural gas extraction
f extraction du gaz naturel
e extracción de gas natural
استخراج الغاز الطبيعى

2355 natural gas liquids
f condensats de gaz naturel
e gas natural licuado
سوائل الغاز الطبيعى

2356 natural hazard
f risque naturel
e riesgo natural
خطر طبيعى

2357 natural monument
f monument naturel
e monumento natural
معلم طبيعى

2358 natural ozone
f ozone naturel
e ozono natural
الأوزون الطبيعى

2359 natural pollutant
f polluant naturel
e contaminante natural
ملوث طبيعى

2360 natural purification
f épuration naturelle
e depuración natural
تنقية طبيعية

2361 natural rain
f pluie normale
e lluvia normal
مطر طبيعى

2362 natural reservoir
f réservoir naturel
e reserva natural
مستودع طبيعى

2363 natural resource account
f compte de ressources naturelles
e cuenta de recursos naturales
حساب الموارد الطبيعية

2364 natural resource accounting
f comptabilité des ressources naturelles
e contabilidad de los recursos naturales
المحاسبة المتعلقة بالموارد الطبيعية

2365 natural resource endowment
f patrimoine naturel
e patrimonio natural
تراث الموارد الطبيعية

2366 natural resources
f ressources naturelles
e recursos naturales
موارد طبيعية

2367 natural sink
f puits naturel
e sumidero natural
بالوعة طبيعية

2368 natural succession
f succession naturelle
e sucesión natural
تسلسل طبيعى

2369 natural-resource base
f ressources naturelles
e recursos naturales
قاعدة الموارد الطبيعية

2370 naturally grown
f obtenu naturellement
e obtenido naturalmente
نابت طبيعياً

2371 naturally-aspirated engine
f moteur à aspiration naturelle
e motor de aspiración natural
محرك يعمل بالسحب الطبيعىللهواء

2372 naturally-occurring repository
f réceptacle géologique naturel
e depósito natural
مكمن طبيعى

2373 naturally-occurring substance
f substance d'origine naturelle
e sustancia natural
مادة طبيعية

2374 nature conservation
f conservation de la nature
e conservación de la naturaleza
صيانة الطبيعة

2375 nature laboratory
f laboratoire de la nature
e laboratorio de la naturaleza
مختبر الطبيعة

2376 nature reserve
f réserve naturelle
e reserva natural
محتجز طبيعى

2377 navigational hazards
f danger de navigation
e peligros para la navegación
مخاطر الملاحة البحرية

2378 near drop-in substitute
f produit n'exigeant que peu d'adaptation
e sucedáneo casi inmediato
بديل يتطلب قليلا من التكييف

2379 near-infrared spectrum
f spectre en proche infrarouge
e espectro del infrarrojo cercano
الطيف دون الأحمر الأدنى

2380 near-shore fishing
f pêche côtière
e pesca costera
صيد الأسماك بالقرب من الشواطئ

2381 near-shore pollution
f pollution causée au voisinage des côtes
e contaminación costera
تلوث قريب من الشاطئ

2382 needle-leaved forest
f forêt aciculifoliée
e bosque de coníferas
غابات الأشجار الابرية

2383 neighbourhood improvement schemes
f programmes d'amélioration des quartiers
e planes de mejoras de barrios
مخططات تحسين المناطق المجاورة

2384 nephanalysis
f néphanalyse
e nefoanálisis
تحليل السحب

2385 nesting ground
f aire de nidification
e zona de nidificación
منطقة تعشيش

2386 nesting site
f site de nidification
e lugar de nidificación
موقع تعشيش

2387 net emissions
f émissions nettes
e emisiones netas
صافى الانبعاثات

2388 net polluter
f exportateur net (de polluants)
e contaminador neto
ملوث خالص

2389 net receiver
f importateur net
e importador neto
منطقة تلقى (ملوثات)

2390 net resource depletion
f épuisement net des ressources
e agotamiento neto de los recursos
استنفاد الموارد الصافي

2391 neutralizer
f neutralisant
e neutralizante
معادل

2392 neutrosphere
f neutrosphère
e neutrosfera
الغلاف التعادلى

2393 new and renewable sources of energy
f sources d'énergie nouvelles et renouvelables
e fuentes de energía nuevas y renovables
مصادر الطاقة الجديدة والمتجددة

2394 new communities
f nouvelles communautés
e nuevas comunidades
مجتمعات محلية جديدة

2395 new plant variety
f nouvelle obtention végétale
e nueva variedad vegetal
صنف نباتى جديد

2396 night-time lorry traffic ban
f interdiction faite aux camions de circuler la nuit
e prohibición del tráfico nocturno de camiones
حظر مرور الشاحنات ليلاً

2397 night-time ozone profile
f profil de répartition nocturne d'ozone
e curva de distribución nocturna del ozono
مقطع رأسى ليلى للأوزون

2398 nitrate pollution
f pollution par les nitrates
e contaminación por nitratos
تلوث بالنيترات

2399 nitrates
f nitrates
e nitratos
نيترات

2400 nitrites
f nitrites
e nitritos
حامض النيتروز

2401 nitrocompound
f composé nitré
e compuesto nitrogenado
مركب نيتريتى

2402 nitrogen blanket
f atmosphère d'azote
e atmósfera de nitrógeno
غطاء نيتروجينى

2403 nitrogen catalysis
f catalyse du monoxyde d'azote
e catálisis del nitrógeno
حفز بالنيتروجين

2404 nitrogen chemistry
f chimie de l'azote
e química del nitrógeno
كيمياء النيتروجين

2405 nitrogen compound
f composé azoté
e compuesto de nitrógeno
مركب نيتروجينى

2406 nitrogen cycle
f cycle de l'azote
e ciclo del nitrógeno
دورة النيتروجين

2407 nitrogen fertilizer
f engrais azoté
e fertilizante nitrogenado
سماد نيتروجينى

2408 nitrogen fixation
f fixation de l'azote
e fijación del nitrógeno
تثبيت النيتروجين

2409 nitrogen fixing plants
f plantes vertes fixatrices d'azote
e plantas fijadoras de nitrógeno
نباتات مثبتة للنيتروجين

2410 nitrogen oxides
f oxydes d'azote
e oxidos de nitrógeno
أكاسيد النيتروجين

2411 nitrogen pollution
f pollution par l'azote
e contaminación por compuestos nitrogenados
تلوث بالنيتروجين

2412 nitrogen radicals
f radicaux azotés
e radicales nitrógeno
شقوق النيتروجين

2413 nitrogenous wastes
f déchets azotés
e desechos nitrogenados
نفايات نيتروجينية

2414 nitrosamines
f nitrosamines
e nitrosaminas
نيتروسامين

2415 nitrous fumes
f vapeurs nitreuses
e vapores nitrosos
أبخرة نيتروجينية

2416 no-effect level
f niveau à effet nul
e nivel de efecto nulo
مستوى انعدام التأثير

2417 noise abatement
f lutte anti - bruit
e lucha contra el ruido
مكافحة الضوضاء

2418 noise black spot
f point noir du bruit
e punto negro de ruido
النقطة السوداء للضوضاء

2419 noise limits
f valeurs limites d'émission acoustique
e limites de las emisiones acústicas
الحدود القصوى للضجيج

2420 noise monitoring
f surveillance du bruit
e vigilancia del ruido
رصد الضوضاء

2421 noise pollution
f pollution acoustique
e contaminación acústica
تلوث بالضوضاء

2422 noise pollution level
f niveau de pollution acoustique
e nivel de contaminación acústica
مستوى التلوث الضوضائى

2423 noise screen
f écran anti-bruit
e pantalla insonora
ستارة عزل الصوت

2424 nomads
f nomades
e nómadas
رحل

2425 non-agricultural land
f terre incultivable
e tierras no agrícolas
أراض غير زراعية

2426 non-catalyst car
f voiture non équipée d'un catalyseur
e automóvil sin catalizador
سيارة غير مزودة بحفاز

2427 non-climate factor
f facteur non climatique
e factor no climático
عامل غير مناخى

2428 non-compliance
f inobservation
e incumplimiento
عدم الامتثال

2429 non-consumptive use
f exploitation rationnelle
e explotación racional
استعمال رشيد

2430 non-conventional virus resistance
f résistance induite aux virus
e resistencia adquirida a los virus
مقاومة غير طبيعية للفيروسات

2431 non-culturable micro-organism
f micro-organisme non-cultivable
e microorganismo no cultivable
كائن عضوى دقيق غيرقابل للاستنبات

2432 non-degradable organic compounds
f composés organiques non dégradables
e compuestos orgánicos no degradables
مركبات عضوية غير قابلة للتحلل

2433 non-farm uses of cropland
f terre ayant perdu sa vocation agricole
e aprovechamiento no agrícola de las tierras
استخدام غير زراعى للأراضي

2434 non-forest use
f utilisation non forestière
e aprovechamiento no forestal
استعمال غير حرجى

2435 non-germline cell
f cellule de la lignée somatique
e célula somática
خلية خط غير جرثومى

2436 non-greenhouse gas
f gaz sans effet de serre
e gas sin efecto invernadero
من غير غاز الاحتباس الحرارى

2437 non-indigenous organism
f organisme allogène
e organismo exótico
كائن عضوى غير محلى

2438 non-inert waste
f déchet non inerte
e desecho no inerte
نفايات غير خاملة

2439 non-ionizing radiation
f rayonnement non-ionisant
e radiación no ionizante
إشعاع غير مؤين

2440 non-malignant
f bénin
e benigno
غير خبيث

2441 non-marginal changes
f changements non marginaux
e cambios no marginales
تغيرات غير هامشية

2442 non-marketed good
f produit non marchand
e producto no comercializado
سلع غير مسوقة

2443 non-miscible liquid
f liquide immiscible
e líquido no miscible
سائل غير مخلوط

2444 non-mixing bowl pollutant
f polluant non réactif
e contaminante no reactivo
ملوث غير تفاعلى

2445 non-modified organism
f organisme non modifié
e organismo no modificado
كائن غير محور

2446 non-point source
f source non ponctuelle
e fuente no localizada
مصدر غير محدد

2447 non-polluting energy sources
f sources d'énergie non polluantes
e fuentes de energía no contaminadas
مصادر طاقة غير ملوثة

2448 non-proprietary name
f dénomination commune
e denominación común
إسم غير تجارى

2449 non-recyclable material
f matière non recyclable
e material no reciclable
مادة غير قابلة للتدوير

2450 non-renewable energy resources
f sources d'énergie non renouvelable
e recursos de energía no renovable
مصادر الطاقة غير المتجددة

2451 non-renewable resources
f ressources non- renouvelables
e recursos no-renovables
موارد غير متجددة

2452 non-renewal exploitation
f exploitation destructrice
e explotación no renovable
استغلال غير قابل للتعويض

2453 non-reusable packing
f emballage perdu
e envase sin retorno
تغليف استهلاكى

2454 non-saturated
f non saturé
e no saturado
غير مشبع

2455 non-selective trapping
f piégeage à caractère non sélectif
e captura no selectiva con trampa
الصيد العشوائى بالشراك

2456 non-structural mitigation
f atténuation non structurelle
e mitigación no estructural
تخفيف غير هيكلى

2457 non-sustainable
f non durable
e no sostenible
غير قابل للاستدامة

2458 non-tidal
f non soumis aux marées
e no afectado por la marea
غير متعلق بالمد والجزر

2459 non-timber forest product
f produit forestier non ligneux
e producto forestal no leñoso
منتج حرجى غير خشبى

2460 non-treatable refuse
f refus de traitement
e desecho no tratable
نفايات لا يمكن معالجتها

2461 non-uniform flow
f écoulement varié
e flujo variable
تدفق غير موحد

2462 non-user value
f valeur de non-usage
e valor de no utilización
قيمة غير استعمالية

2463 non-vascular plant
f plante avasculaire
e planta avascular
نبات غير وعائى

2464 non-waste technology
f techniques sans déchets
e tecnología sin desechos
تكنولوجيا عديمة النفايات

2465 non-wood forest products
f produits forestiers non ligneux
e productos forestales no leñosos
منتجات حرجية غير خشبية

2466 North Atlantic drift
f dérive nord-atlantique
e deriva del Atlántico Norte
تيار شمال المحيط الأطلسى

2467 North Pacific drift
f dérive nord-pacifique
e deriva del Pacífico Norte
تيار شمال المحيط الهادئ

2468 not restricted
f non réglementé
e no restringido
غير محظور

2469 not-in-my-back-yard syndrome
f syndrome de rejet
e síndrome de rechazo
ظاهرة إبعاد النفايات

2470 notification
f notification
e notificación
إخطار

2471 noxious emissions
f émissions toxiques
e emisiones tóxicas
انبعاثات ضارة

2472 nuclear accidents
f accidents nucléaires
e accidentes nucleares
حوادث نووية

2473 nuclear energy
f energie nucléaire
e energía nuclear
طاقة نووية

2474 nuclear energy uses
f usages de l'énergie nucléaire
e usos de la energía nuclear
استخدام الطاقة النووية

2475 nuclear fuels
f combustibles nucléaires
e combustibles nucleares
وقود نووى

2476 nuclear power plants
f centrales nucléaires
e plantas nucleares
محطات قوى نووية

2477 nuclear research centres
f centres de recherche nucléaires
e centros de investigación nuclear
مراكز بحوث نووية

2478 nuclear safety
f sécurité nucléaire
e seguridad nuclear
السلامة النووية

2479 nuclear waste disposal site
f décharge nucléaire
e vertedero de desechos nucleares
موقع لتصريف النفايات النووية

2480 nuclear weapons
f armes nucléaires
e armas nucleares
أسلحة نووية

2481 nuclear-free zone
f zone dénucléarisée
e zona desnuclearizada
منطقة خالية من الأسلحة النووية

2482 nucleic acid probe
f sonde d'acide nucléique
e sonda de ácido nucleico
مسار الحمض النووى

2483 nuisance
f nuisance
e molestia
ضرر

2484 nursery area (for fish)
f zone d'alevinage
e criadero
منطقة تفريخ (اسماك)

2485 nutrient (material)
f nutriment
e sustancia nutritiva
مادة مغذية

2486 nutrient depletion
f épuisement des nutriments
e agotamiento de los nutrientes
نفاد المغذيات

2487 nutrient leaching
f lessivage des nutriments
e lavado de nutrientes
رشح المغذيات

2488 nutrient requirement
f besoin en nutriments
e necesidades de nutrientes
الاحتياجات من المغذيات

2489 nutrient salts
f sels nutritifs
e sales nutritivas
الأملاح المغذية

2490 nutrient turnover
f renouvellement des nutriments
e renovación de los elementos nutritivos
دوران المغذيات

2491 nutrient-laden sediment
f alluvion riche en éléments nutritifs
e aluvión rico en nutrientes
راسب غنى بالمغذيات

2492 nutrients
f nutriments
e nutrientes
مغذيات

2493 nutrition
f alimentation
e nutrición
تغذية

2494 nutrition and health care
f nutrition et santé
e nutrición y salud
التغذية والرعاية الصحية

2495 nutritive value of food
f valeur nutritive des aliments
e valor nutritivo de los alimentos
القيمة الغذائية للأغذية

O

2496 oak plantation
f chênaie
e robledal
مزرعة شجر بلوط

2497 obliterative shading
f ombres de camouflage
e gama de coloración de camuflaje
تدرج لونى قاتم

2498 obscuration (of smoke)
f obscurcissement(dûs aux fumées)
e opacidad (del humo)
عتامة (الدخان)

2499 observational needs
f observations nécessaires
e observaciones necesarias
احتياجات المراقبة

2500 observational synthesis
f synthèse fondée sur l'observation
e síntesis basada en la observación
تحليل بالمراقبة

2501 occupational health
f hygiène du travail
e higiene en el lugar de trabajo
صحة مهنية

2502 occupational safety
f sécurité du travail
e seguridad en el lugar de trabajo
سلامة مهنية

2503 ocean basin
f bassin océanique
e cuenca oceánica
حوض محيطى

2504 ocean chemistry
f chimie de la mer
e química marina
كيمياء المحيطات

2505 ocean circulation
f circulation océanique
e circulación oceánica
دوران المحيطات

2506 ocean currents
f courants océaniques
e corrientes oceánicas
تيارات المحيطات

2507 ocean dumping
f immersion de déchets dans l'océan
e inmersión de desechos en el océano
إلقاء فى المحيطات

2508 ocean dynamics
f dynamique des océans
e dinámica de los oceános
ديناميات المحيطات

2509 ocean eddy
f tourbillon en haute mer
e torbellino oceánico
دوامة محيطية

2510 ocean engineering
f génie océanique
e ingeniería oceánica
الهندسة البحرية

2511 ocean environment
f milieu océanique
e medio oceánico
بيئة المحيطات

2512 ocean mapping
f cartographie océanique
e cartografía marina
رسم خرائط المحيطات

2513 ocean pollution
f pollution marine
e contaminación marina
تلوث المحيطات

2514 ocean station
f station océanique
e estación oceánica
محطة بحرية

2515 ocean temperature
f température des océans
e temperatura del océano
درجة حرارة المحيطات

2516 ocean thermal energy conversion
f conversion de l'énergie thermique des mers
e conversión de la energía térmica de los mares
تحويل الطاقة الحرارية للبحار

2517 ocean warming
f échauffement des océans
e calentamiento de los océanos
ارتفاع درجة حرارة المحيطات

2518 ocean water
f eau océanique
e agua oceánica
مياه المحيطات

2519 ocean-atmosphere interaction
f interaction océan-atmosphère
e interacción océano-atmósfera
التفاعل بين المحيط والغلاف الجوى

2520 oceanic absorption
f absorption océanique
e absorción oceánica
امتصاص المحيطات (للطاقة)

2521 oceanography
f océanographie
e oceanografía
الأقيانوغرافيا

2522 oceans
f océans
e océanos
محيطات

2523 octane number
f indice d'octane
e octanaje
رقم الاوكتان

2524 odorant
f odorisant
e odorante
ذو رائحة

2525 odour nuisance
f nuisance olfactive
e molestias producidas por los olores
مضايقة بسبب الروائح

2526 offensive odour
f odeur désagréable
e olor desagradable
رائحة كريهة

2527 off-gases
f effluents gazeux
e efluentes gaseosos
إنبعاثات غازية

2528 off-gases
f effluents gazeux
e efluentes gaseosos
إنبعاثات غازية

2529 offices
f bureaux
e oficinas
مكاتب

2530 off-odour
f odeur atypique
e olor atípico
رائحة غريبة

2531 off-odour
f odeur atypique
e olor atípico
رائحة غريبة

2532 off-peak commuting
f voyages quotidiens au lieu de travail aux heures creuses
e desplazamientos de trabajo en horas de poco tráfico
الذهاب والعودة وقت انفراج الازدحام

2533 off-peak working
f travaux en périodes creuses
e trabajo y movilización fuera de horas críticas
عمل خلال انخفاض ذروة العمل

2534 off-road vehicle
f véhicule tout terrain
e vehículo todo terreno
مركبة صالحة للمناطق الوعرة

2535 off-setting income support
f soutien compensatoire du revenu
e apoyo para la compensación de los ingresos
دعم معوض للدخل

2536 offset
f compensation
e compensación
عوض

2537 off-shore oil drilling
f forage des puits de pétrole offshore
e perforación petrolera costa afuera
تنقيب عن النفط فى عرض البحر

2538 off-shore sewage outfall
f émissaire marin
e emisario marino
مخارج التصريف في البحر

2539 off-stoichiometric combustion
f combustion non stoechiométrique
e combustión no estequiométrica
احتراق غير نقي

2540 off-the-shelf technologies
f technologies disponibles immédiatement
e tecnologías disponibles
تكنولوجيات جاهزة

2541 offshore area
f zone en mer
e zona mar adentro
منطقة بحرية

2542 oil boom
f barrage flottant
e barrera flotante de contención
حاجز عائم

2543 oil collection vessel
f navire de récupération des hydrocarbures
e buque recolector
سفينة جمع النفط المنسكب

2544 oil extraction
f extraction du pétrole
e extracción del petróleo
استخراج الزيت

2545 oil film
f couche d'huile
e película de petróleo
طبقة زيتية رقيقة

2546 oil pollution control
f lutte contre la pollution par les hydrocarbures
e lucha contra la contaminación por petróleo
مكافحة التلوث النفطي

2547 oil pollution damage
f dommages dûs à la pollution par les hydrocarbures
e daños causados por la contaminación por petróleo
أضرار التلوث النفطي

2548 oil recovery
f récuperation d'hydrocarbures
e recuperación de petróleo
إستعادة النفط

2549 oil residue recuperation
f récupération des déchets d'huile
e recuperación de desechos de aceites usados
استعادة بقايا النفط

2550 oil retention barrier
f barrage de retenue
e barrera de contención del petróleo
حاجز لاحتواء النفط

2551 oil seed rape
f colza
e colza
بذر اللفت الزيتي

2552 oil shales
f schistes bitumeux (kérogènes)
e esquistos bituminosos
طفل زيتى

2553 oil slick
f nappe de pétrole brut
e mancha de aceite en el agua
بقعة نفطية

2554 oil spill
f marée noire
e derrame de petróleo
انسكاب نفطي

2555 oil spill control
f lutte contre la marée noire
e lucha contra los derrames de petróleo
مكافحة الانسكابات النفطية

2556 oil tankers
f pétroliers
e buques petroleros
ناقلات النفط

2557 oil-fired
f au fioul
e de fuelóleo
يعمل بالنفط

2558 oils
f huiles et pétroles
e petróleos minerales
زيوت

2559 old secondary forest
f forêts secondaires anciennes
e antiguo bosque de segunda generación
غابة ثانوية قديمة

2560 oligophotic zone
f zone oligophotique
e zona oligofótica
منطقة قليلة الضوء

2561 oligotrophic
f oligotrophe
e oligotrófico
عديم المغذيات

2562 oligotrophic lakes
f lacs oligotrophes
e lagos oligotróficos
بحيرات نادرة المغذيات النباتية

2563 on-farm testing (of technologies)
f essai sur le terrain des techniques d'exploitation agricole
e ensayo sobre el terreno
اختبار (التكنولوجيات) في المزارع

2564 on-shore wind
f vent d'afflux
e viento de mar
ريح شاطئية

2565 on-site
f in situ
e in situ
في الموقع

2566 once-through
f en circuit ouvert
e de paso único
بدون إعادة تدوير

2567 onchocerciasis
f onchocercose
e oncocercosis
داء المذنبات الملتحية

2568 opacity
f opacité
e opacidad
لا انفادية

2569 open burning
f brûlage à l'air libre
e combustión al aire libre
حرق في الهواء الطلق

2570 open dump
f décharge brute
e verterdero abierto
مدفن قمامة مكشوف

2571 open forest
f forêt claire
e bosque abierto
غابة غير ممتلئة

2572 open pasture land
f parcours herbeux
e pastizal
أراض الرعي المفتوحة

2573 open spaces
f espaces verts
e espacios abiertos
مساحات مفتوحة

2574 open tree formation
f forêt claire
e masa forestal clara
أشجار متباعدة

2575 open woodland
f forêt claire
e bosque claro
حرجة غير ممتلئة

2576 open-cup test
f essai en creuset ouvert
e prueba en crisol abierto
إختبار الكأس المفتوحة

2577 opencast mining
f extraction à ciel ouvert
e minería a cielo abierto
تعدين سطحي

2578 opening of woodlands for agriculture
f mise en culture des terres boisées
e tala de bosques para la agricultura
فتح الأراضي المحرجة للزراعة

2579 operational area
f lieu de l'intervention
e zona de operaciones
منطقة عمليات

2580 optical depth
f profondeur optique
e profundidad óptica
العمق الضوئي

2581 optical window
f fenêtre optique
e ventana óptica
نافذة ضوئية

2582 optically active gas
f gaz optiquement actif
e gas ópticamente activo
غاز نشط ضوئيا

2583 option price
f prix d'option
e precio de opción
سعر الخيار

2584 option value
f valeur d'option
e valor de opción
قيمة الخيار

2585 orbiting satellite
f satellite à défilement
e satélite en órbita
ساتل سيار

2586 ore deposits
f gisement de minerais
e yacimientos de minerales
رواسب الخام

2587 organic
f matière organique
e orgánico
عضوي

2588 organic chemistry
f chimie organique
e química orgánica
كيمياء عضوية

2589 organic content of soils
f teneur en matières organiques des sols
e contenido orgánico de los suelos
المحتوى العضوي للتربة

2590 organic farming
f agriculture biologique
e agricultura orgánica
زراعة باستخدام أسمدة طبيعية

2591 organic farming
f culture organique
e agricultura orgánica
زراعة عضوية

2592 organic manuring
f engrais organiques
e abonos orgánicos
تسميد عضوي

2593 organic pollutants
f polluants organiques
e contaminantes orgánicos
ملوثات عضوية

2594 organic solvents
f solvants organiques
e disolventes orgánicos
مذيبات عضوية

2595 organic substances
f substances organiques
e sustancias orgánicas
مواد عضوية

2596 organochloride
f organochloré
e organoclorado
كلوريد عضوي

2597 organochlorine compound
f composé organochloré
e compuesto organoclorado
مركب عضوي كلوري

2598 organohalogen compound
f composé organohalogéné
e compuesto organohalogenado
مركب عضوي هالوجيني

2599 organometallic compound
f composé organométallique
e compuesto organometálico
مركب عضوي فلزي

2600 organometallic species
f organométalliques
e especies organometálicas
مجموعات عضوية فلزية

2601 organophosphates
f composés organophosphorés
e organofosfatos
مركبات فوسفاتية عضوية

2602 organophosphorous compounds
f composés organo-phosphoriques
e compuestos organofosfóricos
مركبات الفسفور العضوى

2603 organosilicon compounds
f composées organosiliciés
e compuestos orgánicos de silicio
مركبات السيليكون العضوية

2604 organotin
f composé organique de l'étain
e compuesto organoestánnico
مادة عضوية قصديرية

2605 organotin compounds
f composés organostanniques
e compuestos organoestánnicos
مركبات عضوية قصديرية

2606 orthotropic
f orthotrope
e ortótropo
مستقيم الساق

2607 outbreak area
f aire grégarigène
e foco de invasión
منطقة تفشي

2608 outbreak control
f lutte contre les poussées épidémiques
e lucha contra los brotes epidémicos
مكافحة تفشي وباء

2609 outbreak of fire
f début d'incendie
e foco de incendio
اندلاع النيران

2610 outdoor recreation resources
f ressources récréatives de plein air
e recursos recreativos al aire libre
الموارد الترويحية في الهوا الطلق

2611 outfall
f émissaire
e emisario
مخارج تصريف

2612 outfall concentration
f teneur au point de rejet
e concentración de descargas
التركيز عند مخارج التصريف

2613 outflow
f effluent
e efluente
دفق

2614 outlet
f exutoire
e salida
مخرج

2615 outright ban
f interdiction catégorique
e prohibición total
حظر تام

2616 outside air
f air ambiant
e aire exterior
هواء طلق

2617 over-exploitation
f exploitation excessive
e sobreexplotación
إسراف في الاستغلال

2618 over-exposed individual
f personne ayant subi une irradiation excessive
e persona sobreexpuesta
شخص متعرض لجرعة زائدة

2619 over-felling
f exploitation abusive
e tala excesiva
الإسراف في قطع الأشجار

2620 overbank flooding
f inondation lors de crues
e rebose
فيضان (نهري)

2621 overburdening (of the water supply)
f surcharge (du système d'approvisionnement en eau)
e sobrecarga (del abastecimiento de agua)
السحب الزائد(من إمدادات المياه)

2622 overcropping
f surexploitation des terres
e sobrecultivo
الزراعة المفرطة

2623 overcrowding
f surpopulation
e apiñamiento humano
ازدحام

2624 overcutting
f surexploitation des peuplements forestiers
e sobrecorte
الإفراط في قطع الغابات

2625 overfertilization
f surfertilisation
e uso excesivo de fertilizantes
تسميد مفرط

2626 overfire air
f air additionnel pour la combustion
e aire adicional
هواء إضافي لمساعدة الاحتراق

2627 overfishing
f surpêche
e sobrepesca
الصيد المفرط

2628 overflow filter
f filtre déversant
e filtro de rebosadero
مرشح الطفح

2629 overgrazing
f surpâturage
e pastoreo excesivo
الرعي المفرط

2630 overintensive agriculture
f agriculture superintensive
e agricultura superintensiva
زراعة مفرطة

2631 overland flow
f écoulement de surface
e escorrentía superficial
إنسياب المياه على السطح

2632 overloading
f surcharge
e sobrecarga
تحميل زائد

2633 overmature stand
f peuplement suranné
e masa decadente
حرجة تجاوزت حد العمر

2634 overpollution
f surpollution
e contaminación excesiva
تلوث مفرط

2635 overprediction
f surestimation
e sobreestimación
إفراط في التنبؤ

2636 overriding priority
f priorité absolue
e prioridad absoluta
أولوية مطلقة

2637 overstocking
f surpâturage
e sobrecarga de ganado
تخزين مفرط

2638 overstorey layer
f étage supérieur
e piso superior
الطبقة العليا (لغابة)

2639 overtopping frequency
f fréquence de franchissement par la mer
e frecuencia de rebosamiento
تواتر ارتفاع مياه البحر فوق الحواجز

2640 overutilization
f surexploitation
e aprovechamiento excesivo
الإستغلال المفرط

2641 overwintering plant
f espèce pérenne
e planta vivaz
نبات دائم الخضرة

2642 oxidation
f oxydation
e oxidación
أكسدة

2643 oxidation catalyst
f catalyseur d'oxydation
e catalizador de oxidación
حفاز للأكسدة

2644 oxidation pond
f étang d'oxydation
e fosa séptica de oxidación
حوض أكسدة

2645 oxidation rate
f vitesse d'oxydation
e velocidad de oxidación
معدل الأكسدة

2646 oxidation tank
f bac d'oxydation
e cuba de oxidación
خزان أكسدة

2647 oxidizing air
f air comburant
e aire oxidante
الهواء المؤكسد

2648 oxidizing solids
f matières comburantes
e sólidos oxidantes
مواد صلبة مؤكسدة

2649 oxygen
f oxygène
e oxígeno
اكسجين

2650 oxygen depletion
f déperdition d'oxygène
e agotamiento del oxígeno
نفاد الأكسجين

2651 oxygen sink
f réservoir d'oxygène
e depósito de oxígeno
بالوعة أكسجين

2652 oxygen-consuming capacity
f oxydabilité
e capacidad de consumo de oxígeno
قدرة على استهلاك الأكسجين

2653 oxygenated fuels
f carburants oxygénés
e combustibles oxigenados
وقود مؤكسد

2654 ozone's vertical distribution
f répartition verticale de l'ozone
e distribución vertical del ozono
التوزع العمودي للأوزون

2655 ozone balance
f équilibre de l'ozone
e equilibrio del ozono
توازن الأوزون

2656 ozone behaviour
f comportement de l'ozone
e comportamiento del ozono
سلوك الأوزون

2657 ozone budget
f bilan d'ozone
e balance del ozono
ميزانية الأوزون

2658 ozone changes
f modification de la couche d'ozone
e alteración del ozono
التغيرات في الأوزون

2659 ozone cloud
f champ d'ozone stratosphérique
e nube de ozono estratosférica
سحابة اوزونية

2660 ozone column
f colonne d'ozone
e columna de ozono
عمود الأوزون

2661 ozone content
f teneur en ozone
e contenido de ozono
محتوى الأوزون

2662 ozone control
f régulation de l'ozone
e control del ozono
مراقبة الأوزون

2663 ozone cycle
f cycle de l'ozone
e ciclo del ozono
دورة الأوزون

2664 ozone decline
f appauvrissement de la couche d'ozone
e reducción del ozono
انخفاض كمية الأوزون

2665 ozone decrease
f diminution de l'ozone
e disminución del ozono
نقص الأوزون

2666 ozone depleter
f substance menaçant l'ozone
e sustancia nociva para el ozono
مادة مستنفدة للأوزون

2667 ozone depleting
f appauvrissement en ozone
e agotamiento del ozono
استنفاد الأوزون

2668 ozone destruction
f décomposition de l'ozone
e descomposición del ozono
تدمير الأوزون

2669 ozone dilution effect
f effet de dilution de l'ozone
e efecto de dilución del ozono
تأثير تخفيف الأوزون

2670 ozone episode
f épisode d'ozone
e episodio de ozono
فترة ازدياد تركيز الأوزون

2671 ozone field
f champ d'ozone
e campo de ozono
حقل اوزوني

2672 ozone hole
f trou (de la couche) d'ozone
e agujero en la capa de ozono
ثقب الأوزون

2673 ozone increase
f augmentation d'ozone
e aumento del ozono
زيادة في الأوزون

2674 ozone layer
f couche d'ozone
e capa de ozono
طبقة الاوزون

2675 ozone layer degrading
f dégradation de la couche d'ozone
e degradación de la capa de ozono
تآكل طبقة الأوزون

2676 ozone layer depletion
f appauvrissement de la couche d'ozone
e agotamiento de la capa de ozono
استنفاد طبقة الأوزون

2677 ozone layer loss
f appauvrissement de la couche d'ozone
e reducción de la capa de ozono
فقد طبقة الأوزون

2678 ozone loss
f perte d'ozone
e pérdida de ozono
نقص الأوزون

2679 ozone map
f carte des quantités d'ozone
e mapa del ozono
خريطة الأوزون

2680 ozone maximum
f concentration maximum d'ozone
e concentración máxima de ozono
التركيزات القصوى للأوزون

2681 ozone minimum
f concentration minimum d'ozone
e concentración mínima de ozono
التركيزات الدنيا للأوزون

2682 ozone model
f modèle de l'ozone
e modelo del ozono
النموذج الاوزوني

2683 ozone modification
f altération de l'ozone
e alteración del ozono
تغير الأوزون

2684 ozone monitoring
f surveillance de l'ozone
e vigilancia del ozono
رصد الأوزون

2685 ozone monitoring station
f station de surveillance de l'ozone
e estación de vigilancia del ozono
محطة رصد الأوزون

2686 ozone observational record
f registre des données sur l'ozone
e registro de datos sobre el ozono
سجل بيانات الأوزون

2687 ozone observing station
f station d'observation de l'ozone
e estación de observación del ozono
محطة قياس الأوزون

2688 ozone partial pressure
f pression partielle en ozone
e presión parcial del ozono
الضغط الجزئى الأوزونى

2689 ozone pollution
f pollution par l'ozone
e contaminación por ozono
تلوث الأوزون

2690 ozone precursors
f précurseurs de l'ozone
e precursores del ozono
سلائف الأوزون

2691 ozone producer
f producteur d'ozone
e productor de ozono
مولد للأوزون

2692 ozone profile
f courbe de répartition d'ozone
e curva de distribución del ozono
مقطع رأسي لتركيز الأوزون

2693 ozone regime
f régime de l'ozone
e régimen del ozono
نظام الأوزون

2694 ozone science
f ozonologie
e ciencia del ozono
علم الأوزون

2695 ozone shield
f bouclier d'ozone
e capa de ozono
درع الأوزون

2696 ozone sink
f puits d'ozone
e sumidero del ozono
بالوعة أوزون

2697 ozone sonde
f sonde pour l'ozone
e ozonosonda
مسبار الأوزون

2698 ozone sounding
f sondage(s) de l'ozone
e sondeo del ozono
سبر الأوزون

2699 ozone trend(s)
f évolution de l'ozone
e evolución del ozono
اتجاهات الأوزون

2700 ozone value
f valeur de l'ozone
e valor del ozono
كمية الأوزون

2701 ozone-damaging emissions
f émissions nocives pour l'ozone
e emisiones perjudiciales para el ozono
انبعاثات ضارة بالأوزون

2702 ozone-depleting potential
f potentiel d'appauvrissement de la couche d'ozone
e potencial de agotamiento del ozono
طاقة استنفاد الأوزون

2703 ozone-forming species
f composé ozonogène
e compuesto ozonogénico
مجموعات مكونة للأوزون

2704 ozone-friendly
f qui préserve la couche d'ozone
e inocuo para el ozono
غير ضار بالأوزون

2705 ozone-friendly technology
f techniques protectrices de la couche d'ozone
e tecnología inocua para el ozono
تكنولوجيا غير ضارة بالأوزون

2706 ozone-poor air
f air pauvre en ozone
e aire pobre en ozono
هواء فقير بالأوزون

2707 ozone-protecting technology
f techniques assurant la protection de l'ozone
e tecnología para la protección del ozono
تكنولوجيا حامية لطبقة الأوزون

2708 ozone-rich air
f air riche en ozone
e aire rico en ozono
هواء غني بالأوزون

2709 ozonization
f ozonation
e ozonización
المعالجة بالأوزون

2710 ozonizer
f ozoniseur
e ozonador
مولد الأوزون

2711 ozonocide
f ozonocide
e ozonocida
مبيد للأوزون

2712 ozonolysis
f ozonolyse
e ozonolisis
تحليل بالأوزون

2713 ozonometer
f ozonomètre
e ozonómetro
مقياس الأوزون

2714 ozonoscope
f ozonoscope
e ozonoscopio
مكشاف الأوزون

2715 ozonosphere
f ozonosphère
e ozonosfera
طبقة الأوزون

P

2716 pack ice
f banquise
e banquisa
كتلة جليدية طافية

2717 package of measures
f train de mesures
e conjunto de medidas
مجموعة تدابير

2718 packaged
f sous emballage
e embalado
مغلف

2719 packaging
f emballage
e embalaje
تغليف

2720 packaging waste
f déchets d'emballage(s)
e desechos de embalaje
نفايات التغليف

2721 packed refuse
f déchets compactés
e desechos compactados
نفايات مضغوطة

2722 packed tower
f tour à garnitures
e torre rellena
برج ترشيح بالضغط

2723 packer truck
f benne à compression
e camión compactador
شاحنة قمامة ضاغطة

2724 paint solvent
f solvant pour peinture
e disolvente de pintura
مذيب الدهان

2725 paints
f peintures
e pinturas
طلاء

2726 pairing
f appariement
e apareamiento
مزاوجة

2727 palaeoclimatology
f paléoclimatologie
e paleoclimatología
علم مناخ ما قبل التاريخ

2728 palaeoenvironment
f paléoenvironnement
e paleoentorno
البيئة القديمة

2729 par-release test
f test de prédissémination
e ensayo previo a la liberación
الاختبار السابق للإطلاق

2730 parameterization of radiation
f paramétrage du rayonnement
e parameterización de la radiación
وضع معايير للإشعاع

2731 parasites
f parasites
e parásitos
طفيليات

2732 parental organism
f organisme parental
e organismo parental
الكائن العضوي الأصل

2733 partially halogenated
f partiellement halogéné
e parcialmente halogenado
مهلجن جزئيا

2734 partially halogenated hydrocarbon
f hydrocarbure partiellement halogéné
e hidrocarburo parcialmente halogenado
هيدروكربون مهلجن جزئيا

2735 participatory research
f recherche participative
e investigación en participación
بحث قائم على المشاركة

2736 particle-size distribution
f granulometrie
e granulometría
توزع حجمي للجسيمات

2737 particulate control
f lutte contre les émissions de particules
e reducción de la emisión de particulas
منع انبعاث الجسيمات

2738 particulate emission
f émission de particules
e emisión de partículas
انبعاث الجسيمات

2739 particulate filter
f filtre à particules
e filtro de partículas
مرشح جسيمات

2740 particulate loading(s)
f concentration en particules
e concentración de partículas
حمل الجسيمات

2741 particulate matter
f matières particulaires
e materia particulada
جسيمات

2742 particulate precipitator
f séparateur des particules
e precipitador de partículas
جهاز فصل الجسيمات

2743 particulate trap
f piège à particules
e trampa para partículas
مصيدة جسيمات

2744 partition chromatography
f chromatographie de partage
e cromatografia de partición
الفصل اللوني بالتقسيم

2745 passive sensor
f détecteur passif
e detector pasivo
جهاز إستشعار سلبي

2746 passive solar heating system
f dispositif héliothermique passif
e sistema de calentamiento solar pasivo
نظام تسخين شمسي سلبي

2747 passive user
f utilisateur passif
e usuario pasivo
مستخدم غير مستهلك

2748 passport information
f données d'enregistremant
e datos de registro
بيانات التعريف

2749 pastoralist
f pasteur
e pastor
راع

2750 patent culture
f culture destinée à être brevetée
e cultivo patentable
استنبات مسجل ببراءة

2751 patenting
f brevetabilité
e patentabilidad
تسجيل البراءات

2752 pathogen-free drinking water
f eau potable saine
e agua potable salubre
مياه شرب صحية

2753 pathogenic organisms
f organismes pathogènes
e organismos patógenos
كائنات حية مسببة للأمراض

2754 pathogenicity
f pathogénicité
e patogenicidad
قابلية التسبب بالمرض

2755 pathology report
f rapport pathologique
e informe patológico
تقرير عن مسببات الأمراض

2756 pathway of a pollutant
f cheminement d'un polluant
e recorrido de un contaminante
مسار الملوث

2757 pattern of energy consumption
f structure de la consommation d'énergie
e estructura del consumo de energía
نمط استهلاك الطاقة

2758 pattern of events
f réseau d'incidents
e pauta de incidentes
نمط (تتابع) الأحداث

2759 pattern of temperatures
f diagramme thermique
e diagrama térmico
نمط درجات الحرارة

2760 patterns of urban growth
f modèles de croissance urbaine
e tendencia del crecimiento urbano
أنماط النمو الحضرى

2761 paucity of natural reserves
f modicité des réserves naturelles
e escasez de reservas naturales
ندرة المحميات الطبيعية

2762 peak chlorine loading
f concentration maximale de chlore
e contenido máximo de cloro
حمل الكلور الأقصى

2763 peak concentration
f concentration de pointe
e concentración máxima
تركيز أقصى

2764 peak flow
f débit de pointe
e caudal máximo
تدفق أقصى

2765 peat
f tourbe
e turba
فحم المستنقعات

2766 peatlands
f tourbières
e turbera
تربة خثية

2767 pedosphere
f pédosphère
e pedosfera
الغلاف الترابى

2768 pelagic drift-net fishing
f pêche hauturière aux filets dérivants
e pesca de altura con redes de enmalle y de deriva
صيد الأسماك بشباك الجر

2769 pelagic pollution
f pollution pélagique
e contaminación pelágica
تلوث أعماق البحار

2770 pelagic zone
f zone pélagique
e zona pelágica
منطقة أعالي البحار

2771 pelleted catalyst
f catalyseur granulé
e catalizador granulado
حفاز مزود بكريات

2772 penalties for environmental damage
f sanctions pour atteinte à l'environnement
e sanciones por daños ecológicos
جزاءات على الضرر البيئى

2773 penetration time
f temps d'imprégnation
e tiempo de penetración
فترة التغلغل

2774 people-centered development
f développement centré sur la population
e desarrollo centrado en la población
تنمية مركزة على السكان

2775 perchlorinated
f perchloré
e perclorado
مشبع بالكلور

2776 percolation
f percolation
e percolación
ترشح

2777 percutaneous
f par voie cutanée
e a través de la piel
عن طريق الجلد

2778 perfluorinated
f perfluoré
e perfluorado
مشبع بالفلور

2779 performance bond
f caution de bon fonctionnement
e fianza de cumplimiento
سند ضمان حسن الأداء

2780 performance characteristics
f caractères de performance
e características funcionales
خصائص الأداء

2781 performance of the adsorbent
f efficacité de l'adsorbant
e rendimiento del adsorbente
فعالية المادة الممتصة

2782 performance resins
f résines de haute résistance
e resinas de alto rendimiento
راتنجات جيدة الأداء

2783 performance trial
f essai de rendement
e prueba de rendimiento
اختبار الأداء

2784 perhalogenated
f perhalogéné
e perhalogenado
كامل الهلجنة

2785 peri-urban settlement
f établissement péri-urbain
e asentamiento periurbano
مستوطنة قريبة من منطقة حضرية

2786 period of drought
f période de sécheresse
e periodo de sequaي
فترة الجفاف

2787 permafrost
f permagel
e permafrost
ارض دائمة التجمد

2788 permanent crops
f cultures vivaces
e cultivos perennes
محاصيل دائمة

2789 permanent meadow
f prairie permanente
e prado permanente
مرج دائم

2790 permanent plot
f parcelle permanente
e parcela permanente
أرض مخصصة للمراقبة الدائمة

2791 permanent waste storage
f stockage définitif des déchets
e almacenamiento permanente de los desechos
الخزن الدائم للنفايات

2792 permissible exposure limit
f limite d'exposition admissible
e límite de exposición admisible
حد التعرض المسموح به

2793 persistence of pesticides
f persistance des pesticides
e persistencia de plaguicidas
مداومة مبيدات الآفات

2794 persistent organic pollutants
f polluants organiques persistants
e contaminantes orgánicos persistentes
ملوثات عضوية مداومة

2795 pest outbreak
f invasion de parasites
e invasión de parásitos
تفشي الآفات

2796 pesticide control standards
f normes de contrôle des pesticides
e normas de control de plaguicidas
معايير مكافحة مبيدات الآفات

2797 pesticide pathways
f chaîne des réactions associées à la biodégradation des pesticides
e cadena de reacciones asociadas con la biodegradación de plaguicidas
تتابع تفاعلات مبيدات الآفات

2798 pesticide tolerance
f limite de tolérance des pesticides
e límite de tolerancia de plaguicidas
مبيدات الآفات المسموح بها

2799 pesticides
f pesticides
e plaguicidas e insecticidas
مبيدات الآفات

2800 petrol
f pétrole
e petróleo
بنزين

2801 petrol fumes
f vapeurs d'essence
e vapores de gasolina
أبخرة البنزين

2802 petroleum geology
f géologie pétrolière
e geología del petróleo
جيولوجيا النفط

2803 petroleum refining
f raffinage du pétrole
e refinación del petróleo
تكرير النفط

2804 petroleum resources conservation
f conservation des ressources
e conservación de los recursos de petróleo
صيانة الموارد النفطية

2805 pH
f pH
e pH
درجة حموضة

2806 pH changes
f variations du pH
e cambios del pH
التغيرات فى درجة الحموضة

2807 pH control
f régulation du pH
e regulación del pH
تنظيم درجة الحموضة

2808 pH decrease
f diminution du pH
e disminución del pH
نقص درجة الحموضة

2809 pharmaceutical wastes
f déchets pharmaceutiques
e desechos farmacéuticos
نفايات صيدلانية

2810 pharmacology
f pharmacologie
e farmacología
علم العقاقير

2811 phase down
f réduction progressive
e reducción progresiva
خفض تدريجي

2812 phasing out
f élimination progressive
e supresión progresiva
إلغاء تدريجي

2813 phenols
f phénols
e fenoles
فينولات

2814 phenotype
f phénotype
e fenotipo
مظهر موروث

2815 pheromones
f phéromones
e feromonas
روائح الجاذبية الجنسية

2816 philoprogenitive
f génitophile
e filoprogenitivo
كثير الانسال

2817 phosphate rock
f phosphate naturel
e fosfato mineral
فوسفات صخري

2818 phosphates
f phosphates
e fosfatos
فوسفات

2819 photo-oxidant
f oxydant photochimique
e fotooxidante
مؤكسد ضوئي

2820 photoautotroph
f phototrophe
e fotoautótrofo
ذاتي التغذية بالضوء

2821 photocatalysis
f photocatalyse
e fotocatálisis
حفز بالضوء

2822 photochemical agents
f agents photochimiques
e agentes fotoquímicos
عوامل ضوئية كيميائية

2823 photochemical dispersion model
f modèle de dispersion photochimique
e modelo de dispersión fotoquímica
نموذج للتشتت الكيميائي الضوئي

2824 photochemical effects
f effets photochimiques
e efectos fotoquímicos
آثار ضوئية كيميائية

2825 photochemical oxidants
f oxydants photochimiques
e oxidantes fotoquímicos
مؤكسدات كيميائية ضوئية

2826 photochemical pollution
f pollution (d'origine) photochimique
e contaminación fotoquímica
تلوث كيميائي ضوئي

2827 photochemical smog
f brouillard photooxydant
e niebla fotoquímica
ضباب دخاني كيميائي ضوئي

2828 photochemically reactive
f photochimiquement réactif
e fotoquímicamente reactivo
متفاعل بالطرق الكيميائية الضوئية

2829 photochemistry of ozone
f photochimie de l'ozone
e fotoquímica del ozono
الكيمياء الضوئية للأوزون

2830 photodissociation
f photodissociation
e fotodisociación
تفكك بالضوء

2831 photogrammetry
f photogrammétrie
e fotogrametría
تصوير مساحى

2832 photolysis
f photolyse
e fotolisis
تحلل ضوئي

2833 photomap
f photocarte
e fotomapa
خريطة فوتوغرافية

2834 photoperiod
f photopériode
e fotoperíodo
فترة ضوئية

2835 photoperiodism
f photopériodisme
e fotoperiodicidad
خاصية الاستجابة للفترة الضوئية

2836 photoprotection
f photoprotection
e fotoprotección
حماية ضوئية

2837 photoreactivation
f photoréactivation
e fotoreactivación
إعادة تنشيط الضوء

2838 photorespiration
f photorespiration
e fotorespiración
تنفس محفز بالضوء

2839 photosensitization
f photosensibilisation
e fotosensibilización
حساسية الجلد للضوء

2840 photosensitizer
f photosensibilisateur
e fotosensibilizador
مادة مسببة لحساسية الجلد للضوء

2841 photosynthesis
f photosynthèse
e fotosíntesis
تمثيل ضوئى

2842 photosynthetic rate
f vitesse de la photosynthèse
e velocidad de la fotosíntesis
معدل التمثيل الضوئي

2843 physical alterations
f modifications physiques
e alteraciones físicas
تغيرات مادية

2844 physical amplification factor
f coefficient physique d'incidence
e coeficiente de amplificación física
عامل التضخيم الفيزيائي

2845 physical chemist
f physico-chimiste
e fisico químico
أخصائي كيمياء فيزيائية

2846 physical chemistry
f chimie physique; physico-chimie
e fisicoquímica
الكيمياء الفيزيائية

2847 physical climate system
f système climatique physique
e sistema climático fisico
نظام المناخ الطبيعي

2848 physical environment
f cadre de vie
e entorno fisico
البيئة المادية

2849 physical oceanography
f océanographie physique
e oceanografía física
اقيانوغرافيا فيزيائية

2850 physical planning
f aménagement du territoire
e ordenación territorial
التخطيط العمراني

2851 physical resource accounts
f comptes de ressources naturelles
e cuenta de recursos naturales
حسابات الموارد الطبيعية

2852 physico-chemical processes
f processus physico-chimiques
e procesos fisicoquímicos
عمليات فيزيائية – كيميائية

2853 phytocenosis
f phytocénose
e fitocenosis
موئل نباتي

2854 phytocide
f phytocide
e fitocida
مبيد نباتي

2855 phytomass
f phytomasse
e fitomasa
كتلة نباتية

2856 phytoplankton
f phytoplancton
e fitoplancton
عوالق نباتية

2857 phytoplankton bloom
f prolifération phytoplanctonique
e proliferación de fitoplancton
تكاثر العوالق النباتية

2858 phytotoxic
f phytotoxique
e fitotóxico
سام للنبات

2859 pick-up point
f point de prise en charge
e punto de recolección
نقطة جمع القمامة

2860 pigmentophore cell
f cellule pigmentophore
e célula cromatófora
خلية ملونة

2861 pilot projects
f projets pilotes
e proyectos pilotos
مشروعات تجريبية

2862 pioneer species
f essence pionnière
e especie pionera
أنواع رائدة

2863 pipelines
f oléoduc
e oleoductos
خطوط أنابيب

2864 plagiotropic
f plagiotrope
e plagiótropo
أفقي النمو

2865 planetary boundary layer
f couche limite planétaire
e capa límite planetaria
الطبقة الحدية حول الأرض

2866 planetary jet
f courant-jet planétaire
e corriente en chorro planetaria
تيار هوائي سريع دائم

2867 plankton
f plancton
e plancton
عوالق

2868 plankton multiplier
f multiplicateur planctonique
e multiplicador planctónico
التأثير المضاعف للعوالق

2869 plankton recorder
f enregistreur planctonique
e registrador de plancton
كمية العوالق

2870 planned release
f libération volontaire
e liberación voluntaria
إطلاق منظم

2871 planned urban development
f aménagement urbain planifié
e desarrollo urbano planificado
تنمية حضرية مخططة

2872 plant environment
f milieu végétal
e medio vegetal
البيئة النباتية

2873 plant breeding
f phytogénétique
e mejoramiento de plantas
استنبات

2874 plant cover
f couverture végétale
e cubierta vegetal
الغطاء النباتي

2875 plant damage
f lésion(s) des plantes
e daños sufridos por las plantas
ضرر يصيب النبات

2876 plant diseases
f maladies des plantes
e enfermedades de las plantas
أمراض النبات

2877 plant ecology
f écologie végétale
e ecología vegetal
الايكولوجيا النباتية

2878 plant genetic resource
f ressource phytogénétique
e recurso fitogenético
مورد وراثي نباتي

2879 plant genetics
f phytogénétique
e fitogenética
جينات النباتات

2880 plant introduction
f introduction végétale
e introducción de especies vegetales
إدخال النبات

2881 plant life
f flore
e flora
الحياة النباتية

2882 plant nutrition
f phytotrophie
e nutrición de las plantas
تغذية النبات

2883 plant physiology
f physiologie végétale
e fisiología vegetal
فسيولوجيا النبات

2884 plant protection
f protection phytosanitaire
e protección fitosanitaria
حماية النبات

2885 plant protein
f protéine végétale
e proteína vegetal
بروتين نباتي

2886 plant waste
f déchets verts
e desperdicios vegetales
نفايات نباتية

2887 plant-breeders rights
f droits des phytogénéticiens
e derechos de los mejoradores de plantas
حقوق أصحاب المشاتل

2888 planting material
f plant
e plantones
اشتال للزرع

2889 planting media
f terreau
e mantillo
التربة

2890 plastic fabric
f tissu en matière plastique
e tejido plástico
نسيج لدائني

2891 plastic foam
f mousse plastique
e plástico expansible
رغوة لدائنية

2892 plastic wastes
f déchets plastiques
e desechos plásticos
نفايات اللدائن

2893 plasticizer
f plastifiant
e plastificante
ملدن

2894 plate filter
f filtre-presse
e filtro-prensa
مرشح قرصي

2895 playgrounds
f terrains de jeu
e campos de recreo
ملاعب

2896 pleiotropy
f pléiotropie
e pleitropía
تعدد الخصائص

2897 pleomorphism
f pléomorphisme
e pleomorfismo
تعدد الأشكال

2898 plume
f panache
e penacho
عمود دخان

2899 plume centreline
f axe du panache
e eje del penacho
محور عمود الدخان

2900 plume spread
f étalement du panache
e difusión del penacho
انتشار عمود الدخان

2901 plume travel time
f temps de trajet du panache
e tiempo de propagación del penacho
الوقت اللازم لتبدد عمود الدخان

2902 poaching
f braconnage
e caza ilegal
صيد غير مشروع

2903 pocket beach
f crique
e playuela
خليج صغير

2904 point pollution
f pollution ponctuelle
e contaminación puntual
تلوث ثابت المصدر

2905 point source
f source ponctuelle
e fuente individual
مصدر ثابت

2906 poison
f substance toxique
e sustancia tóxica
سم

2907 poisonous gas
f gaz toxique
e gas tóxico
غاز سام

2908 poisonous plant
f plante vénéneuse
e planta venenosa
نبات سام

2909 polar aurora
f aurore polaire
e aurora polar
الشفق القطبي

2910 polar cap
f calotte polaire
e casquete polar
الغطاء الجليدى القطبي

2911 polar ecosystems
f écosystèmes polaires
e ecosistemas polares
النظم الايكولوجية القطبية

2912 polar front
f front polaire
e frente polar
جبهة قطبية

2913 polar ice sheet
f calotte glaciaire polaire
e manto de hielo polar
غطاء جليدي قطبي

2914 polar ozone
f ozone polaire
e ozono polar
أوزون قطبي

2915 polar stratospheric cloud
f nuage stratosphérique polaire
e nube estratosférica polar
سحاب ستراتوسفيري قطبي

2916 polar vortex
f vortex polaire
e vórtice polar
دوامة قطبية

2917 policy planning
f politique de la planification
e planificación de la política
تخطيط السياسة العامة

2918 pollen analysis
f analyse de pollen
e análisis de polen
تحليل حبوب اللقاح

2919 pollutant
f polluant
e contaminante
ملوث

2920 pollutant analysis
f analyse des polluants
e análisis de contaminates
تحليل الملوثات

2921 pollutant distribution
f distribution des polluants
e distribución de contaminantes
توزيع الملوثات

2922 pollutant effects
f effets de polluants
e efectos de contaminantes
آثار الملوثات

2923 pollutant flux
f courant des polluants
e flujo de contaminantes
تدفق الملوثات

2924 pollutant levels
f niveaux des polluants
e niveles de contaminantes
مستويات الملوثات

2925 pollutant pathways
f cheminement des polluants
e trayectoria de contaminantes
تتابع تفاعلات الملوثات

2926 pollutant reduction
f réduction des émissions de polluants
e reducción de la contaminación
تخفيض الملوثات

2927 pollutant removal
f élimination des polluants
e eliminación de contaminantes
إزالة الملوثات

2928 pollutant source identification
f identification des sources de pollution
e identificación de la fuente de contaminación
تحديد مصدر الملوث

2929 polluter pays principle
f principe du pollueur payeur
e principio de quien contamina paga
مبدأ الغرم على الملوث

2930 pollution
f pollution
e contaminación
التلوث

2931 pollution abatement
f réduction de la pollution
e reducción de la contaminación
مكافحة التلوث

2932 pollution abatement equipment
f moyens de réduction de la pollution
e equipo de reducción de la contaminación
معدات مكافحة التلوث

2933 pollution control
f lutte antipollution
e lucha contra la contaminación
مكافحة التلوث

2934 pollution control industry
f écoindustrie
e ecoindustria
صناعة مكافحة التلوث

2935 pollution control measures
f mesures antipollution
e medidas de lucha contra la contaminación
تدابير مكافحة التلوث

2936 pollution control policy
f politique antipollution
e política de lucha contra la contaminación
سياسة مكافحة التلوث

2937 pollution control regulations
f réglementation de la lutte contre la pollution
e reglamentación para el control de la contaminación
قواعد مكافحة التلوث

2938 pollution control ship
f navire dépollueur
e buque descontaminador
سفينة لمكافحة التلوث

2939 pollution control technology
f écotechnologie
e tecnología para la reducción de la contaminación
تكنولوجيا مكافحة التلوث

2940 pollution controls
f réglementation de la dépollution
e reglamentación de la lucha contra la contaminación
ضوابط التلويث

2941 pollution criteria
f critères de pollution
e criterios de contaminación
معايير التلوث

2942 pollution from non-point sources
f pollution diffuse
e contaminación difusa
التلوث من مصادر غير ثابتة

2943 pollution incident
f pollution accidentelle
e caso de contaminación
حادث تلوث

2944 pollution level
f degré de pollution
e grado de contaminación
مستوى التلوث

2945 pollution liabilities
f responsabilité de la pollution
e resposabilidad por contaminación
مسؤوليات عن التلوث

2946 pollution load
f charge polluante
e carga de contaminación
حمل التلوث

2947 pollution monitoring
f surveillance de la pollution
e vigilancia de la contaminación
رصد التلوث

2948 pollution norms
f normes de pollution
e normas sobre contaminación
أنماط التلوث

2949 pollution rent
f redevance de pollution
e canon de contaminación
غرامة التلويث

2950 pollution risk
f risque de pollution
e riesgos de la contaminación
خطر التلوث

2951 pollution sources
f sources de pollution
e fuentes de contaminación
مصادر التلوث

2952 pollution taxes
f redevances pour pollution
e impuesto a la contaminación
ضرائب على التلويث

2953 pollution-immune
f non affecté par la pollution
e inmune a la contaminación
عديم التأثر بالتلوث

2954 pollution-prone
f prédisposé à la pollution
e propenso a la contaminación
معرض للتلوث

2955 polychlorinated biphenyls
f biphényls polychlorés
e bifeniles policlorados
ثنائيات الفنيل المتعدد الكلور

2956 polyhalogenated
f polyhalogéné
e polihalogenado
متعدد الهالوجين

2957 polymer wastes
f déchets de polymères
e desechos polímeros
نفايات البلمرات

2958 pond life
f faune des eaux stagnantes
e fauna y flora palustres
أحياء المياه الراكدة

2959 ponds tailings
f écoulement des étangs
e estanques de decantación
نفايات البرك

2960 poplar plantation
f peupleraie
e alameda
مزرعة شجر الحور

2961 population ecology
f écologie des populations
e ecología de las poblaciones
ايكولوجيا السكان

2962 pore water
f eau interstitielle
e agua intersticial
المياه المسامية

2963 ports
f ports
e puertos
موانئ

2964 posidonia meadows
f prairies de posidonies
e campos de posidonias
مروج مغمورة

2965 positive environmental qualities
f qualités antipollution
e cualidades ambientales positivas
الخصائص البيئية الإيجابية

2966 post-disaster relief
f secours après la catastrophe
e socorro después de los desastres
الإغاثة بعد الكارثة

2967 post-disaster syndrome
f syndrome post-catastrophe
e síndrome de secuela del desastre
أعراض ما بعد الكارثة

2968 post-effect of drought
f arrière-effet de le la sécheresse
e secuela de la sequía
الآثار اللاحقة للجفاف

2969 potent
f actif
e potente
فعال

2970 potential difference
f différence de potentiel
e diferencia de potencial
فرق الجهد

2971 potentially harmful substance
f substance potentiellement nocive
e sustancia potencialmente peligrosa
مادة محتمله الضرر

2972 potentially polluting activity
f activité potentiellement polluante
e actividad potencialmente contaminante
نشاط محتمل التلويث

2973 poultry farming
f aviculture
e avicultura
تربية الدواجن

2974 poverty alleviation
f lutte contre la pauvreté
e mitigación de la pobreza
تخفيف حدة الفقر

2975 poverty eradication
f suppression de la pauvreté
e erradicación de la pobreza
القضاء على الفقر

2976 poverty line
f seuil de pauvreté
e umbral de pobreza
خط الفقر

2977 poverty redressal
f élimination de la pauvreté
e eliminación de la pobreza
علاج الفقر

2978 powder waste
f déchets pulvérulents
e desechos pulverulentos
نفايات مسحوقة

2979 powdery substance
f matière pulvérulente
e sustancia pulverulenta
مادة مسحوقة

2980 power gas
f gaz combustible
e gas combustible
غاز وقودي

2981 power line
f ligne de force
e línea de transporte de energía
خط كهرباء

2982 power outage
f coupure de courant
e corte de energía
انقطاع التيار

2983 prawn farming
f élevage de crevettes
e cría de camarones
تربية الاربيان

2984 pre-acidification alkalinity
f alcalinité initiale
e alcalinidad inicial
القلوية السابقة للتحمض

2985 pre-disaster activities
f mesures de prévention des catastrophes
e actividades en previsión de desastres
الأنشطة السابقة لوقوع الكارثة

2986 pre-investment feasibility study
f étude de faisabilité de préinvestissement
e estudio de preinversión
دراسة الجدوى السابقة للاستثمار

2987 pre-sorting
f tri préalable (des déchets)
e separación previa
تصنيف أولى

2988 pre-spill planning
f planification des interventions en cas de déversement
e planificación previa a los derrames
تخطيط سابق لحدوث الانسكاب

2989 precautionary approach
f démarche fondée sur le principe de précaution
e enfoque basado en el principio de precaución
نهج وقائي

2990 precautionary principle
f principe de précaution
e principio de precaución
مبدأ وقائي

2991 precipitate
f précipité
e precipitado
راسب

2992 precipitation enhancement
f intensifications des précipitations
e aumento de la precipitación
زيادة تساقط المطر

2993 precipitator
f dépoussiéreur
e precipitador
جهاز ترسيب

2994 precollection
f précollecte
e prerrecolección
تصنيف النفايات السابق لجمعها

2995 precursor pollutant
f polluant précurseur
e contaminante precursor
مادة مشكلة لملوث

2996 predispose to damage
f rendre plus vulnérable aux atteintes
e predisponer a los efectos nocivos
عرضة للضرر

2997 prefabricated buildings
f bâtiments préfabriqués
e edificios prefabricados
مبانى جاهزة للتركيب

2998 preparedness
f préparation
e preparación
التأهب

2999 preparedness planing
f planification préalable
e planificación de la preparación
تخطيط التأهب

3000 prescrubber
f dispositif de prélavage
e dispositivo de prelavado
جهاز غسل أولى

3001 preserve
f zone protégée
e zona reservada
منطقة محمية

3002 pressure groups
f groupes de pression
e grupos de presión
مجموعات ضغط

3003 pressure on forest resources
f surexploitation des ressources forestières
e sobreexplotación de los recursos forestales
الضغط على موارد الغابات

3004 pretreatment
f prétraitement
e pretratamiento
معالجة أولية

3005 prevailing conditions
f conditions existantes
e condiciones reinantes
الأحوال السائدة

3006 prevailing wind
f vent dominant
e viento dominante
الرياح السائدة

3007 preventive maintenance
f entretien préventif
e mantenimiento preventivo
صيانة وقائية

3008 pricing policies of resources
f politique des prix des ressources naturelles
e política de precios de los recursos naturales
سياسات تسعير الموارد

3009 prilling
f granulation
e granulación
تحبب

3010 prills
f gouttelettes solidifiées
e glóbulos de metal en las escorias
حبيبات

3011 primary colour
f couleur primaire
e color primario
لون أساسي

3012 primary consumers
f consommateurs primaires
e consumidores primarios
مستهلكون أولون

3013 primary cosmic radiation
f rayonnement cosmique primaire
e radiación cósmica primaria
أشعة كونية أولية

3014 primary energy
f énergie primaire
e energía primaria
طاقة أولية

3015 primary environmental care
f protection environnementale primaire
e protección ambiental primaria
العناية البيئية الأولية

3016 primary forest
f forêt vierge
e bosque virgen
غابة بكر

3017 primary formation
f terrain primitif
e formación primaria
تكوين أولى

3018 primary impact area
f zone d'impact primaire
e zona de impacto primario
منطقة التأثير الأولي

3019 primary pollutant
f polluant primaire
e contaminante primario
ملوث أولى

3020 primary product
f produit primaire
e producto primario
منتج أولى

3021 primary soil
f sol autochtone
e suelo autóctono
تربة أصلية

3022 primary treatment
f premier traitement
e tratamiento primario
معالجة أولية

3023 primates
f primates
e primates
رئيسيات

3024 principle of prior informed consent
f principe du consentement préalable
e principio del consentimiento fundamentado previo
مبدأ الموافقة المسبقة عن علم

3025 printing industry
f industrie de l'imprimerie
e industria de las artes gráficas
صناعة الطباعة

3026 prior assessment
f évaluation préalable
e evaluación previa
تقييم مسبق

3027 prior informed consent
f consentement basé sur l'information préalable
e consentimiento notificado previo
موافقة مسبقة عن علم

3028 prior notification for hazardous waste transport
f notification préalable aux tranports des déchets dagereux
e notificación previa para transporte de desechos peligrosos
إخطار مسبق لنقل نفايات خطرة

3029 priority chemical
f produit chimique d'intérêt prioritaire
e producto químico prioritario
مواد كيميائية ذات أولوية

3030 pristine area
f zone vierge
e zona virgen
منطقة بكر

3031 probe
f sonde
e sonda
مسبار

3032 problem site
f point noir
e punto negro
موقع مشاكل

3033 problem waste
f déchets encombrants
e desechos problemáticos
نفايات يصعب التخلص منها

3034 process industry
f industrie de transformation
e industria de transformación
صناعة التحويل

3035 process waste
f déchets de production
e desechos de fabricación
نفايات الإنتاج

3036 processing
f traitement
e tratamiento
معالجة

3037 product charges
f redevances sur produits
e impuesto a los productos
رسوم إنتاج

3038 product labeling
f étiquetage des produits
e etiquetado de productos
وضع العلامات على المنتجات

3039 product stewardship
f bonne gestion des produits
e administración de los productos
الإدارة الجيدة للمنتجات

3040 prokaryotic micro-organisms
f micro-organismes procaryotes
e microorganismos procarióticos
كائنات مجهرية بحرية بدائية النواة

3041 propagator
f propagateur
e propagador
جهاز استنبات

3042 propellant (gas)
f propulseur
e gas propulsor
غاز دافع

3043 proprietary rights
f droits de propriété
e derecho de propiedad
حقوق الملكية

3044 protected areas
f zones protégées
e regiones protegidas
مناطق محمية

3045 protected landscape area
f zone de paysage protégé
e zona de paisaje protegido
منطقة مناظر طبيعية محمية

3046 protected species
f espèces protégées
e especies protegidas
أنواع محمية

3047 protected taxa
f taxons protégés
e taxones protegidos
أنواع محمية

3048 protection forest
f forêt de protection
e bosque de protección
غابة حماية

3049 protein balance
f équilibre protéique
e equilibrio proteínico
توازن بروتيني

3050 proteinase inhibitor
f inhibiteur de la protéinase
e inhibidor de la proteinasa
مانع الأنزيم البروتيني

3051 protoplasm fusion
f fusion protoplasmique
e fusión protoplásmica
اندماج بروتوبلازمي

3052 protozoa
f protozoaires
e protozoos
بروتوزوا

3053 protozoal disease
f maladie transmise par les protozoaires
e enfermedad transmitida por protozoarios
مرض تنقله الكائنات الأولية

3054 proximate cause
f cause immédiate
e causa inmediata
السبب المباشر

3055 proxy data
f données indirectes
e datos indirectos
بيانات غير مباشرة

3056 psychrometer
f psychromètre
e psicrómetro
مقياس الجفاف

3057 public access to land
f accès public aux terres
e acceso público a la tierra
حصول الجمهور علىالأرض

3058 public awareness raising
f sensibilisation de la population
e sensibilización de la población
زيادة الوعي العام

3059 public gardens
f jardins publics
e jardines públicos
حدائق عامة

3060 public health
f hygiène publique
e salud pública
الصحة العامة

3061 public land
f terre domaniale
e terreno de dominio público
أراض عامة

3062 public parks
f parcs publics
e parques y jardines públicos
متنزهات عامة

3063 public right to know
f droit du public à l'information
e derecho del público a la información
حق الجمهور في المعرفة

3064 public services
f services publics
e servicios públicos
خدمات عامة

3065 public utilities
f services publics
e servicios públicos
مرافق عامة

3066 public warning
f mise en garde de la population
e aviso a la población
تحذير عام

3067 publicly-owned forest
f forêt publique
e bosque de propriedad pública
غابة حكومية

3068 puff
f bouffée
e bocanada
نفثة

3069 pulp industry
f industrie de la pâte et du papier
e industria de la pulpa y papel
صناعة الورق

3070 pulpwood
f bois à pâte
e madera para pasta
عجين لب الخشب

3071 pulverization
f broyage
e pulverización
سحق

3072 pulverizer
f unité de broyage
e pulverizador
جهاز سحق

3073 pump lift drainage
f drainage par pompage
e drenaje por bombeo
الصرف بالضخ

3074 pumping station
f station de pompage
e estación de bombeo
محطة ضخ

3075 pungent odour
f odeur âcre
e olor acre
رائحة حادة

3076 pure rain
f pluie normale
e lluvia normal
مطر عادي

3077 purification
f épuration
e depuración
تنقية

3078 **purified water**
f eau épurée
e agua depurada
ماء منقى

3079 **purity**
f pureté
e pureza
نقاء

3080 **purpose-built vehicle**
f véhicule spécialisé
e vehículo especializado
مركبة لغرض خاص

Q

3081 quality assurance programme
f programme d'assurance de la qualité
e programa de garantía de la calidad
برنامج ضمان النوعية

3082 quality control
f contrôle de qualité
e control de calidad
رقابة النوعية

3083 quality objectives
f objectifs de qualité
e objetivos de calidad
اهداف النوعية

3084 quality standards
f normes de qualités
e normas de calidad
معايير النوعية

3085 quarry reclamation
f réaménagement de carrière
e recuperación de canteras
استصلاح المحاجر

3086 quarrying
f exploitation de carrières
e explotación de canteras
إستغلال المحاجر

3087 quench layer
f calamine
e calamina
طبقة الرواسب الكربونية

3088 quench tank
f ballon de noyage
e tanque de enfriamiento
صهريج التبريد

3089 quick-growing tree
f arbre à croissance rapide
e árbol de crecimiento rápido
شجرة سريعة النمو

3090 quiet sun
f soleil calme
e sol en calma
شمس هادئة

R

3091 race relations
f relations inter-raciales
e relaciones interraciales
علاقات الأجناس

3092 radiation standards
f normes de radioprotection
e normas de protección radiológica
معايير الاشعاع

3093 radiation accident
f irradiation accidentelle
e irradiación accidental
حادث اشعاعي

3094 radiation balance
f bilan des radiations
e radiación total resultante
ميزانية الاشعاع

3095 radiation damage
f dommages dûs au rayonnement
e daños por irradiación
ضرر اشعاعي

3096 radiation data
f données sur les radiations
e datos radiométricos
بيانات اشعاعية

3097 radiation effects
f effets des radiations
e efectos de la radiación
آثار الإشعاع

3098 radiation emergency
f situation d'urgence radiologique
e situación de emergencia radiológica
طارئ اشعاعي

3099 radiation level
f intensité du rayonnement
e intensidad de la radiación
مستوى الاشعاع

3100 radiation medicine
f radiologie médicale
e medicina de radiación
الطب الاشعاعي

3101 radiation monitoring
f surveillance des radiations
e vigilancia de la radiación
رصد الإشعاع

3102 radiation protection
f protection contre les radiations
e protección contra las radiaciones
وقاية من الإشعاع

3103 radiation sickness
f maladies dues aux radiations
e enfermedad de las radiaciones
أمراض التعرض للإشعاع

3104 radioactive contamination
f contamination radioactive
e contaminación radioactiva
تلوث إشعاعى

3105 radioactive substances
f substances radioactives
e sustancias radioactivas
مواد مشعة

3106 radioactive tracer techniques
f techniques de traçage radioactif
e técnicas de trazadores radioactivos
تقنيات عناصر تتبع الإشعاع

3107 radioactive waste management
f gestion des déchets radioactifs
e manejo de desechos radioactivos
إدارة النفايات المشعة

3108 radioactivity
f radioactivité
e radioactividad
إشعاع

3109 radiobiology
f radiobiologie
e radiobiología
البيولوجيا الاشعاعية

3110 radioecology
f radioécologie
e radioecología
الايكولوجيا الاشعاعية

3111 radiological safety
f sécurité radiologique
e seguridad radiológica
سلامة اشعاعية

3112 radionuclide
f radionucléide
e radionúclido
نويدة مشعة

3113 radiosonde
f radiosonde
e radiosonda
مسبار لاسلكي

3114 radiowind
f radiovent
e radio viento
رياح اشعاعية

3115 radon
f radon
e radón
رادون

3116 railway transport
f transport ferroviaire
e transporte ferroviario
نقل بالسكك الحديدية

3117 rain forest
f forêt ombrophile
e selva (pluvial)
غابة مطيرة

3118 rain recorder
f pluviographe
e pluviógrafo
مسجل الامطار

3119 rain(-)fed agriculture
f agriculture pluviale
e agricultura pluvial
زراعة بعلية

3120 rain(-)out
f incorporation dans les nuages suivie de précipitation
e arrastre producido por la lluvia
غسل مطري

3121 rain-free period
f période sèche
e período seco
فترة جفاف

3122 rain-producing system
f système pluvigène
e sistema pluvígeno
نظام توليد الامطار

3123 rainfall
f chute de pluie
e lluvia
سقوط الامطار

3124 rainfall amount
f hauteur de pluie
e altura de (la) precipitación
كمية الامطار

3125 rainfall regime
f régime pluviométrique
e régimen pluviométrico
نظام سقوط الامطار

3126 rainfall season
f saison des pluies
e estación de las lluvias
فصل الامطار

3127 raingauge
f pluviomètre
e pluviómetro
مقياس الامطار

3128 ram jet
f statoréacteur
e estatorreactor
محرك نفاث ضغاطي

3129 ranching
f élevage
e ganadería
تربية الماشية في المزارع

3130 range finding
f télémétrie
e telemetría
تعيين المدى

3131 rangeland
f pâturages extensifs
e pastizal
مرعى

3132 rapidly developing disasters
f catastrophes à évolution rapide
e desastres de evolución rápida
كوارث سريع التكون

3133 rare taxa
f espèces rares
e taxones raros
انواع نادرة

3134 rate of climate change
f rythme des changements climatiques
e ritmo de los cambios climáticos
معدل التغير المناخي

3135 rate of discharge
f débit de rejet
e velocidad de descarga
معدل التصريف

3136 rate of emission
f débit d'émission
e velocidad de emisión
معدل الانبعاث

3137 rationalization of hunting crops
f rationalisation des tableaux de chasse
e racionalización de los cupos de caza
ترشيد صيد الحيوانات

3138 ravine erosion
f ravinage
e abarrancamiento
تآكل الاودية الصغيرة

3139 raw radiance
f luminance énergétique brute
e radiancia bruta
اشعاعية اولية

3140 raw refuse
f déchets bruts
e desechos brutos
قمامة غير معالجة

3141 raw sewage
f eaux d'égout non traitées
e aguas residuales sin tratar
مياه المجارير غير المعالجة

3142 raw water
f eau non traitée
e agua sin depurar
مياه غير معالجة

3143 re-engineering
f conversion
e reconversión
اعادة هندسة

3144 re-establishment of trees
f reconstitution des forêts
e reforestación
اعادة التحريج

3145 reaction time
f temps de réaction
e tiempo de respuesta
زمن التفاعل

3146 reactive gas
f gaz réactif
e gas reactante
غاز متفاعل

3147 reagent
f réactif
e reactivo
عامل التفاعل

3148 reawakening (of a volcano)
f réveil (d'un volcan)
e reactivación (de un volcán)
استيقاظ البركان

3149 receiving country
f pays récepteur
e país receptor
بلد متلق

3150 receiving environment
f environnement receveur
e medio receptor
بيئة متلقية

3151 receiving waters
f eaux réceptrices
e aguas receptoras
مياه متلقية

3152 receptor area
f région réceptrice
e zona receptora
منطقة استقبال

3153 recession
f tarissement
e recesión
انحسار

3154 recession discharge
f débit de tarissement
e caudal de recesión
تصريف انحساري

3155 recharge area
f zone d'alimentation
e zona de alimentación
منطقة تغذية

3156 recharge of aquifer
f alimentation d'une nappe souterraine
e alimentación de un acuífero
اعادة تغذية طبقة المياه الجوفية

3157 recipient organism
f organisme receveur
e organismo receptor
كائن متلق

3158 recirculation (of gas)
f recyclage (des gaz)
e recirculación
اعادة تدوير (غاز)

3159 reclaimable waste
f déchets à récupérer
e desechos aprovechables
نفايات قابلة للاستيرداد

3160 reclaimed quarry
f carrière réhabilitée
e cantera recuperada
محجر مستصلح

3161 recombinant antigen
f antigène de recombinaison
e antígeno recombinante
مولد المضاد المؤتلف

3162 recombinant DNA technology
f techniques de recombinaison de l'ADN
e tecnología de recombinación genética del ADN
تكنولوجيا توليف الحمض الخلوي الصبغي

3163 recommended exposure limit
f limite d'exposition recommandée
e límite de exposición recomendada
حد التعرض المسموح به

3164 record flood
f inondation sans précédent
e inundación sin precedentes
فيضان قياسي

3165 recording rain gauge
f pluviographe
e pluviógrafo
مقياس مطر مسجل

3166 recovery of higly degraded lands
f régénération de terres extrêmement dégradées
e mejoramiento de tierras muy degradadas
اصلاح الاراضي المتدهورة جدا

3167 recovery of service station vapour
f récupération des gaz des stations service
e recuperación de los gases de las estaciones de servicio
استرداد ابخرة محطات البنزين

3168 recreation
f loisirs
e recreación
ترويح

3169 recreational area
f aire de loisirs
e lugar de recreo
منطقة ترويح

3170 recruits
f recrues
e nuevos individuos
إفراخ السمك

3171 recycled materials
f matières recyclées
e materiales reciclados
مواد معاد دورانها

3172 recyling
f recyclage
e reciclado
اعادة تدوير

3173 red tide
f marée rouge
e marea roja
المد الاحمر

3174 reduced nitrogen species
f composé nitreux réduits
e compuestos de nitrogenados reducidos
فصائل النيتروجين المختزلة

3175 reducing agent
f agent réducteur
e agente reductor
عامل مختزل

3176 refining
f affinage
e depuración
تكرير

3177 reforestation
f reforestation
e repoblación forestal
إعادة التحرج

3178 refractories
f matériaux réfractaires
e materiales refractarios
مواد صامدة للصهر

3179 refrigerant fluid
f fluide réfrigérant
e fluido refrigerante
سائل مبرد

3180 refrigeration capacity
f puissance frigorifique utile
e capacidad refrigerante
قدرة التبريد

3181 refugees
f réfugiés
e refugiados
لاجئون

3182 refuse cart
f bac roulant
e carretilla para basuras
عربة جمع القمامة

3183 refuse chute
f vide-ordures
e conducto para basuras
مسقط القمامة

3184 refuse collector
f benne à ordures
e contenedor para la recolección de basuras
جامع القمامة

3185 refuse receptacle
f contenant pour déchets
e receptáculo para desechos
وعاء القمامة

3186 refuse reclamation
f récupération des déchets
e recuperación de basuras
استرجاع النفايات

3187 refuse treatment
f traitement des déchets
e tratamiento de desechos
معالجة القمامة

3188 refuse truck
f camion benne
e camión de basuras
شاحنة القمامة

3189 refuse-derived fuel
f combustible dérivé de déchets
e combustible obtenido de desechos
وقود مستخلص من الفضلات

3190 regeneration
f régénération
e regeneración
تجديد

3191 regeneration cycle
f cycle de sylvigénèse
e ciclo de regeneración
دورة التجديد

3192 regenerative capability
f reconstituabilité
e capacidad de regeneración
القدرة على التجديد

3193 regenerative heat exchanger
f échangeur de chaleur récupérateur
e termorrecuperador
مبادل حرارة بالاسترجاع

3194 regional conventions
f conventions régionales
e convenios regionales
اتفاقيات إقليمية

3195 regional planning
f planification régionale
e planificación regional
تخطيط إقليمى

3196 regular gas
f essence ordinaire
e gasolina normal
بنزين عادي

3197 regulated flow
f débit régularisé
e corriente regulada
تدفق منظم

3198 regulatory control
f contrôle législatif
e control reglamentario
رقابة ناظمة

3199 rehabilitative measures
f mesures de relèvement
e medidas de readaptación
تدابير اصلاحية

3200 rehousing
f relogement
e realojamiento en nuevas viviendas
إعادة إسكان

3201 release
f fuites
e descargas accidentales
إطلاق

3202 release a vaccine
f commercialiser un vaccin
e comercializar una vacuna
اذن باستخدام لقاح

3203 release of engineered micro-organisms in the environment
f libération de micro-organismes modifiés dans l'environnement
e liberación de microorganismos modificados en el medio ambiente
اطلاق كائنات مجهرية محورة في البيئة

3204 release of heat
f rejet thermique
e liberación de calor
اطلاق الحرارة

3205 release site
f site de dissémination
e lugar de liberación
موقع اطلاق

3206 relief manager
f spécialiste des secours
e director de operaciones de socorro
مدير لشؤون الاغاثة

3207 remedial action
f mesures correctives
e medidas correctivas
تدبير علاجي

3208 remobilization of a chemical
f remise en mouvement d'un produit chimique
e removilización de un producto químico
اعادة تنشيط مادة كيميائية

3209 remote sensing
f télédétection
e teledetección
استشعار عن بعد

3210 remote sensing centre
f centre de télédétection
e centro de teledetección
مركز إستشعار عن بعد

3211 remote sensing imagery
f images fournies par la télédétection
e imágenes obtenidas por teleobservación
التصوير بالاستشعار عن بعد

3212 removal efficiency
f rendement de séparation
e rendimiento del separador
كفاءة الفصل

3213 renewable
f renouvelable
e renovable
متجدد

3214 renewable energy sources
f sources d'énergie renouvelable
e fuentes de energía renovable
مصادر الطاقة المتجددة

3215 renewable resources
f ressources renouvelables
e recursos renovables
موارد متجددة

3216 rental housig
f logements à louer
e viviendas de alquiler
إسكان مؤجر

3217 repeating pattern
f cycle
e ciclo
نمط متكرر

3218 replacement chemical
f substance chimique de remplacement
e sustituto químico
مادة كيميائية بديلة

3219 replacement costs
f coûts de remplacement
e costos de sustitución
تكاليف الإحلال

3220 replenishment
f reconstitution
e abastecimiento
تجديد

3221 replicable model
f modèle reproductible
e modelo práctico
نموذج قابل للتكرار

3222 reporting of an incident
f déclaration en cas d'incident
e comunicación de un incidente
ابلاغ بحادث

3223 reproducibility
f reproductibilité
e reproductibilidad
قابلية الانسال

3224 reproductive capacity
f capacité de reproduction
e capacidad de reproducción
قدرة على التكاثر

3225 reproductive failure
f échec de reproduction
e fracaso reproductivo
فشل التكاثر

3226 reproductive manipulation
f manipulations de la reproduction
e manipulaciones de la reproducción
معالجة الأنسال

3227 reproductive potential
f reproductivité
e reproductividad
احتمال اعادة التكاثر

3228 reptiles
f reptiles
e reptiles
زواحف

3229 repugnant substance
f matière répugnante
e sustancia repugnante
مادة كريهة

3230 research ship
f navire océanographique
e buque oceanográfico
سفينة ابحاث

3231 reservoir
f réservoir
e embalse
خزان

3232 reservoir species
f espèce réservoir
e sustancia retentiva
انواع احتياطية

3233 residential areas
f zones résidentielles
e zonas residenciales
مناطق سكانية

3234 residual
f résidu
e residuo
بقايا

3235 residual fuel oil
f fioul résiduel
e fuelóleo residual
زيت وقود متبق

3236 residual soil
f sol résiduel
e suelo residual
تربة موضعية

3237 resilience of nature
f résilience de la nature
e capacidad de recuperación de la naturaleza
قدرة الطبيعة على الانتعاش

3238 resilience under stress
f résistance aux contraintes extérieures
e resistencia a las tensiones
مرونة تحت اجهاد

3239 resistance gene
f gène de résistance
e gen de resistencia
عامل المقارنة الوراثي

3240 resistant to pollution
f tolérant à la pollution
e resistente a la contaminación
مقاوم للتلوث

3241 resolution
f limite de résolution
e resolución
التحليل

3242 resource appraisal
f évaluation des ressources
e evaluación de recursos
تقدير الموارد

3243 resource base
f ressources (disponibles)
e base de recursos
قاعدة موارد

3244 resource conservation
f conservation des ressources
e conservación de recursos
صيانة الموارد

3245 resource gluttony
f consommation gloutonne de ressources
e avidez por los recursos
استهلاك مفرط للموارد

3246 resource pricing
f formation du prix des ressources
e fijación del precio de los recursos
تسعير الموارد

3247 resource recovery
f récupération
e recuperación de recursos
استخلاص الموارد (من النفايات)

3248 resource-intensive industry
f industrie grosse consommatrice de ressources
e industria consumidora de recursos
صناعة كثيفة استهلاك الموارد

3249 resource-saving
f économe en ressources
e ahorrador de recursos
توفير في الموارد

3250 resources management
f gestion des ressources
e manejo de recursos
إدارة الموارد

3251 respirable-sized particulates
f particules de dimensions inhalables
e partículas de tamaño inhalable
جزئيات بأحجام قابلة للاستنشاق

3252 response
f réaction
e reacción
استجابة ؛ تصدي

3253 response plan
f plan d'intervention
e plan de intervención
خطة استجابة

3254 restocking
f repeuplement
e repoblación
تجديد الأرصدة

3255 restoration of a site
f remise en état d'un site
e restauración de una zona
ترميم موقع

3256 restoration of biological diversity
f restauration de la diversité biologique
e restablecimiento de la diversidad biológica
اصلاح التنوع البيولوجي

3257 restoration of fertility
f correction de la fertilité minérale
e restablecimiento de la fertilidad
استعادة خصوبة

3258 restoration of soils
f régénération des sols
e regeneración de los suelos
اصلاح التربة

3259 restored taxa
f espèces rétablies
e especies restablecidas
أنواع مجددة

3260 restriction enzyme
f enzyme de restriction
e enzima de restricción
انزيم الحصر

3261 retarded spark timing
f retard à l'allumage
e encendido retardado
تأخير الشرارة

3262 retrofit techology
f technique d'adaptation antipollution
e tecnología de modernización
تكنولوجيا التعديل

3263 retroviral gene
f gène de rétrovirus
e gen de retrovirus
جينة محفزة للفيروسات

3264 returnable glass container
f verre consigné
e envase de vidrio retornable
اناء زجاجي مرتجع

3265 reusable containers
f conteneurs réutilisables
e contenedores reutilizables
حاويات يعاد استخدامها

3266 revegetation
f revégétation
e revegetación
إعادة التغطية بالنباتات

3267 reverse osmosis
f osmose inverse
e ósmosis invertida
تناضح عكسي

3268 rigid foams
f mousses rigides
e espuma rígida
رغوات صلبة

3269 rime
f givre
e cencellada blanca
جليد حبيبي أبيض

3270 ring
f noyau cycle
e anillo
حلقة

3271 ring closure
f cyclisation
e ciclización
اكمال الحلقة

3272 ring compound
f composé cyclique
e compuesto cíclico
مركب حلقي

3273 riparian country
f pays riverain
e país ribereño
بلد شاطئي

3274 **riparian habitat for fish**
f habitat pour les poissons le long des cours d'eau
e hábitat ribereño de los peces
موئل أسماك نهري

3275 **rising tide**
f marée montante
e flujo de marea
المد

3276 **risk**
f risque
e riesgo
مخاطرة

3277 **risk management**
f gestion des risques
e gestión de (los) riesgos
ادارة المخاطرة

3278 **risk management of organisms**
f gestion des risques liés aux organismes
e gestión de los riesgos de los organismos
ادارة مخاطر الكائنات

3279 **risk mapping**
f cartographie des zones à risques
e cartografía de las zonas de riesgo
رسم خرائط المناطق المعرضة للخطر

3280 **risk reduction**
f réduction des risques
e reducción de los riesgos
الحد من الاخطار

3281 **risk-prone area**
f zone à risque
e zona expuesta a riesgos
منطقة معرضة للخطر

3282 **river basin development**
f mise en valeur des bassins fluviaux
e aprovechamiento de cuencas fluviales
تنمية أحواض الأنهار

3283 **river basins**
f bassins fluviaux
e cuencas fluviales
أحواض أنهار

3284 **river gauge**
f limnimètre
e limnígrafo
مقياس مناسيب مياه الأنهار

3285 **river pollution**
f pollution des rivières
e contaminación fluvial
تلوث الأنهار

3286 **river realignment**
f régularisation des cours d'eau
e regularización de un cauce
تصحيح مجرى النهر

3287 **river records**
f relevés hydrométriques
e registros hidrométricos
قياسات مياه النهر

3288 **river-borne**
f fluviatile
e transportado por río
محمول بالانهار

3289 river-borne sediments
f alluvions disposées par les cours d'eau
e aluviones fluviales
رواسب نهرية

3290 rivers
f rivières
e ríos
أنهار

3291 road construction
f construction des routes
e construcción y mantenimiento de carreteras
إنشاء الطرق

3292 road maintenance
f entretien des routes
e mantenimiento de las carreteras
صيانة الطرق

3293 road safety
f sécurité routière
e seguridad del tráfico
سلامة الطرق

3294 road traffic engineering
f ingénierie de la circulation routière
e ingeniería de tráfico carretero
هندسة الطرق البرية

3295 road traffic fuel
f carburant moteur
e combustible para vehículos viales
وقود محركات المركبات

3296 road transport
f transports routiers
e transporte por carreteras
نقل برى

3297 roads
f routes
e carreteras
طرق برية

3298 roadside planting
f plantation d'arbres en bordure des routes
e plantación de árboles a lo largo de las carreteras
غرس الأشجار على جوانب الطرق

3299 rock phosphate
f phosphate naturel
e fosfato mineral
فوسفات صخري

3300 root activity
f activité radiculaire
e actividad radical
نشاط الجذور

3301 root environment
f environnement radiculaire
e medio radical
بيئة الجذور

3302 root parasit
f parasite des racines
e parásito de las raíces
طفيليات الجذور

3303 root uptake
f absorption radiculaire
e absorción radicular
مص جذري

3304 root-shoot ratio
f rapport appareil radiculaire/appareil foliaire
e relación raiz-vástago
نسبة الجذور/الفروع

3305 rooting medium
f milieu d'enracinement
e medio de enraizamiento
وسط الجذور

3306 rotting garbage
f ordures en décomposition
e basuras en descomposición
قمامة متعفنة

3307 rough fish
f poisson sans valeur
e peces bastos
سمك عديم القيمة

3308 roughness
f rugosité
e rugosidad
اضطراب

3309 roundwood
f bois ronds
e rollizos
خشب مستدير

3310 route
f voie
e vía
مسلك

3311 routine monitoring
f surveillance systématique
e vigilancia sistemática
رصد منتظم

3312 rubber processing
f traitement du caoutchouc
e procesamiento del caucho
تجهيز المطاط

3313 rubber waste
f déchets du caoutchouc
e desechos del caucho
نفايات المطاط

3314 run-off
f ruissellement et infiltrations
e vaciado de represas
جريان المياه السطحي

3315 run-off model
f modèle à ruissellement
e modelo de escorrentía
نموذج جريان سطحي

3316 running water
f eau courante
e agua corriente
مياه جارية

3317 rural areas
f zones rurales
e zonas rurales
مناطق ريفية

3318 rural frontier
f limite des terres rurales
e límite del medio rural
حدود ريفية

3319 rural land
f espace rural
e tierras rurales
أراض ريفية

3320 rural poor
f pauvres en milieu rural
e pobres del medio rural
فقراء الريف

3321 rural water supply
f approvisionnement en eau dans les zones rurales
e abastecimiento de agua en zonas rurales
إمداد بالمياه في الريف

3322 rural woodlots
f peuplements ruraux
e parcelas rurales
حراج ريفية

S

3323 safe and secure housing
f logement sûr
e vivienda segura
مساكن مأمونة ومضمونة

3324 safe development
f développement sans danger
e desarrollo seguro
تنمية سليمة

3325 safe exposure level
f niveau d'exposition sans danger
e nivel de exposición sin peligro
مستوى التعرض المأمون

3326 safe food
f aliments salubres
e alimentos aptos para el consumo
غذاء مأمون

3327 safe for the environment
f sans danger pour l'environnement
e sin peligro para el medio ambiente
غير ضار بالبيئة

3328 safe use
f utilisation sans danger
e seguridad en el uso
استخدام مأمون

339 safe water
f eau salubre;potable
e agua pura;potable
مياه مأمونة

3330 safeguards
f garanties
e salvaguardias
ضمانات

3331 safeguards for the environment
f mesures de protection de l'environnement
e salvaguardias para el medio ambiente
ضمانات للبيئة

3332 safety in biotechnology
f sécurité en biotechnologie
e seguridad en la biotecnología
السلامة فى التكنولوجيا الاحيائية

3333 safety of releases
f sécurité des libérations
e seguridad de las liberaciones
السلامة في الاطلاق

3334 safety requirements
f règles de sécurité
e requisitos de seguridad
متطلبات السلامة

3335 safety standard for buildings
f normes de sécurité des bâtiments
e normas de seguridad para edificios
معايير سلامة المبانى

3336 saline intrusion
f intrusion d'eau salée
e intrusión de agua salada
تسرب المياه المالحة

3337 **salinization**
f salinisation
e salinización
تملح

3338 **salt balance**
f bilan salin
e balance salino
توازن ملحي

3339 **salt meadow**
f pré salé
e prado salino
مرج مالح

3340 **salt residues**
f sels résiduaires
e sales residuales
مخلفات ملحية

3341 **salt water intrusion**
f intrusion d'eau salée
e intrusión de agua salada
تسرب المياه المالحة

3342 **salt water marsh**
f marais salant
e marisma
مستنقع مياه مالحة

3343 **salting-out**
f relargage
e desplazamiento salino
فصل بالتمليح

3344 **sampler**
f appareil de prélèvement
e sacamuestras
جهاز اخذ العينات

3345 **sampling site**
f point d'échantillonnage
e lugar de toma de muestras
موقع اخذ العينات

3346 **sampling techniques**
f techniques d'échantillonnage
e técnicas de muestreo
تقنيات أخذ العينات

3347 **sanctuary**
f réserve (naturelle)
e refugio natural
ملجأ

3348 **sand deposit**
f dépôt de sable
e depósito de arena
رواسب رملية

3349 **sand dune fixation**
f fixation des dunes de sable
e estabilización de dunas
تثبيت الكثبان الرملية

3350 **sand dunes**
f dunes de sable
e dunas
كثبان رملية

3351 **sand extraction**
f extraction du sable
e canteras de arena y grava
استغلال الرمل

3352 **sand flats**
f laisses de sable
e llanos de arena
سهول رملية

3353 sand wind
f vent de sable
e viento arenoso
رياح رملية

3354 sandy soil
f terrain sablonneux
e terreno arenoso
تربة رملية

3355 sanitarian
f technicien de l'assainissement
e técnico sanitario
عامل في مجال النظافة الصحية

3356 sanitary landfill
f décharge contrôlée
e vertedero sanitario
مدفن قمامة صحي

3357 sanitary landfilling
f mise en décharge contrôlée
e vertimiento sanitario
دفن القمامة الصحي

3358 sanitary sewers
f réseau de drainage des eaux usées
e alcantarillado de aguas residuales
مجار صحية

3359 sanitation
f installations sanitaires
e higiene
مرافق صحية

3360 satellite measurements
f mesures satellitaires
e mediciones por satélite
قياسات بواسطة السواتل

3361 satellite observation
f observation satellitaire
e observación desde satélites
الرصد بالسواتل

3362 satellite ozone data
f données satellitaires sur l'ozone
e datos sobre el ozono obtenidos por satélites
بيانات الاوزون المستقاة بالسواتل

3363 satellite sensor
f détecteur satellisé
e sensor de satélite
جهاز استشعار محمول بساتل

3364 satellite system
f système satellitaire
e sistema emplazado en satélites
شبكة سواتل

3365 satellite-borne
f emporté par satellite
e a bordo de satélite
محمول بساتل

3366 saturated aliphatic compound
f composé aliphatique saturé
e compuesto alifático saturado
مركب دهني مشبع

3367 saturated hydrocabon
f hydrocarbure saturé
e hidrocarburo saturado
هيدروكربون مشبع

3368 saturated zone
f zone saturée
e zona saturada
منطقة مشبعة

3369 saturation vapour pressure
f pression de vapeur saturante
e presión de vapor de saturación
ضغط البخار المشبع

3370 sawdust
f sciure
e aserrín
نشارة الخشب

3371 sawnwood
f sciages
e madera aserrada
خشب منشور

3372 scallop culture
f pectiniculture
e pectinicultura
تربية المحار

3373 scattering event single
f cas de diffusion simple
e caso de dispersión simple
حالة تشتت بسيطة

3374 scatterometer
f diffusiomètre
e difusímetro
مقياس التشتت

3375 scavenge oil
f huile de récupération
e aceite de depuración
زيت الكسح

3376 scavenging agent
f épurateur
e agente depurador
عامل التنقية

3377 scavenging tower
f tour d'épuration
e torre de depuración
برج التنقية

3378 schistosomiasis
f schistosomiase
e esquistosomiasis
الشستوزومية

3379 scientific capacity building
f renforcement des capacités scientifiques
e creación de capacidad científica
بناء القدرة العلمية

3380 scrap metals
f déchets métalliques
e chatarra
معادن خردة

3381 screening (1)
f dépistage
e diagnóstico inicial
تمحيص

3382 screening (2)
f protection blindage
e blindaje
حجب

3383 screenings
f déchets de criblage
e desperdicios del cribado
نفايات الغربلة

3384 scrubber
f épurateur
e depurador
جهاز غسل الغاز

3385 scrubber sludge
f boues de lavage
e lodo de lavado
حمأة الغسل

3386 scrubbing
f lavage
e lavado
الغسل

3387 sea (ice) sheet
f banquise
e banquisa
صفيحة (جليدية) بحرية

3388 sea bed
f fonds marins
e fondos marinos
قاع البحر

3389 sea bed exploitation
f exploitation des fonds marins
e explotación de fondos marinos
استغلال قاع البحر

3390 sea bed mining
f exploitation minière des fonds marins
e minería en fondos marinos
تعدين قاع البحر

3391 sea energy
f énergie des mers
e energía del mar
طاقة بحرية

3392 sea farming
f aquaculture marine
e cultivos marinos
تربية الاحياء البحرية

3393 sea floor renewal
f renouvellement des fonds océaniques
e expansión de los fondos marinos
تجدد تضاريس قاع البحر

3394 sea grass bed
f herbier
e lecho de zosteras y algas marinas
طبقة أعشاب بحرية

3395 sea ice
f glace de mer
e hielo marino
جليد عائم

3396 sea level
f niveau de la mer
e nivel del mar
مستوى سطح البحر

3397 sea outfall
f déversoir en mer
e desagüe marino
مخرج تصريف فى البحر

3398 sea ranch
f enclos marin
e reservas de crianza de animales marinos
مرتع بحري

3399 sea turtle
f tortue marine
e tortuga de mar
سلحفاة بحرية

3400 sea-bed insertion
f enfouissement dans le sol sous-marin
e enterramiento en los fondos marinos
اقحام في قاع البحر

3401 sea-level rise
f élévation du niveau de la mer
e aumento del nivel del mar
ارتفاع مستوى البحر

3402 sea-surface wind
f vent de mer en surface
e viento de la superficie del mar
الرياح البحرية السطحية

3403 seafood tainting
f contamination des aliments de mer
e contaminación de los alimentos marinos
تلوث الأغذية البحرية

3404 search and rescue
f recherche et secours
e búsqueda y salvamento
البحث والانقاذ

3405 seascape
f paysage marin
e paisaje marino
منظر بحري

3406 seasonal deviation
f écart saisonnier
e desviación estacional
انحراف فصلي

3407 seasonal field crop
f culture de plein champ saisonnière
e cultivo extensivo estacional
محصول حقلي موسمي

3408 seasonal fluctuation
f variation saisonnière
e variación estacional
تقلبات فصلية

3409 seasonal overbank flooding
f inondations lors de crues saisonnières
e inundaciones por crecidas estacionales
فيضان ضفافي فصلي

3410 seaweed farming
f culture d'algues
e cultivo de algas
زراعة العشب البحري

3411 secondary cosmic radiation
f rayonnement cosmique secondaire
e radiación cósmica secundaria
الاشعاع الكوني الثانوى

3412 secondary energy
f énergie secondaire
e energía secundaria
طاقة ثانوية

3413 secondary forest
f forêt secondaire
e bosque secundario
غابة ثانوية

3414 secondary impact area
f zone d'impact secondaire
e zona de impacto secundario
منطقة اثر ثانوي

3415 secondary pollutant
f polluant secondaire
e contaminante secundario
ملوث ثانوي

3416 secondary raw material
f matière première de récupération
e materia prima de recuperación
مادة خام ثانوية

3417 secondary treatment
f traitement secondaire
e tratamiento secundario
معالجة ثانوية

3418 sectoral assessment
f évaluation sectorielle
e evaluación temática
تقييم قطاعى

3419 sectoralizing of the environment
f sectorialisation de l'environnement
e sectorización del medio ambiente
تقسيم البيئة الى قطاعات

3420 sediment
f sédiment
e sedimento
رواسب

3421 sediment transport
f transport de sédiments
e transporte de sedimentos
انتقال الرواسب

3422 sedimentary basins
f bassins sédimentaires
e cuencas sedimentarias
أحواض الترسيب

3423 sedimentation
f sédimentation
e sedimentación
ترسب

3424 sedimentation rate
f taux de sédimentation
e tasa de sedimentación
معدل الترسيب

3425 sedimentation tank
f bassin de décantation
e estanque de decantación
حوض الترسيب

3426 seed bank
f banque de semences
e banco de semillas
مصرف بذور

3427 seeded pasture
f prairie artificielle
e pradera cultivada
مرعى اصطناعي

3428 seedling
f plant
e plántula
شتلة

3429 seepage
f suintement
e filtración
نشع

3430 seismic activity
f activités sismiques
e actividad sísmica
نشاط زلزالي

3431 seismic monitoring
f surveillance des activités sismiques
e vigilancia sísmica
رصد الزلازل

3432 seismic sea waves
f raz de marée sismiques
e marejadas sísmicas
أمواج بحرية بفعل الزلازل

3433 selective breeding of animals
f croisement sélectif des animaux
e cría selectiva de animales
تكاثر انتقائي للحيوانات

3434 selective breeding of plants
f croisement sélectif des végétaux
e cruce selectivo de plantas
تكاثر انتقائي للنباتات

3435 selective catalytic reduction
f réduction catalytique sélective
e reducción catalítica selectiva
اختزال حفزي انتقائي

3436 selective collection
f collecte sélective
e recolección selectiva
جمع انتقائي

3437 self-accelerating decomposition
f décomposition auto-accélérée
e descomposición autoacelerada
تحلل ذاتي التسارع

3438 self-desliming
f auto-curage de la vase
e autoeliminación del légamo
النزع الذاتي للطين

3439 self-destroying
f autodégradable
e autodegradable
ذاتي التحلل

3440 self-draining pipe
f conduite à purge gravitaire
e tubo de desagüe automático
انبوب تصريف بالجاذبية

3441 self-help
f auto-assistance
e autoayuda
مساعدة ذاتية

3442 self-help programmes
f programmes d'auto-assistance
e programas de autoayuda
برامج المساعدة الذاتية

3443 semi-arid land ecosystems
f ecosystème des terres semi-arides
e ecosistemas de tierras semiáridas
النظم الايكولوجية للأراضي شبه القاحلة

3444 semi-captive raising of animals
f élevage en semi-liberté
e cría de animales en semicautividad
تربية الحيوانات شبه الحبيسة

3445 semi-enclosed sea
f mer semi-fermée
e mar semicerrado
بحر شبه مغلق

3446 semi-permanent settlement
f semi-sédentarisation
e asentamiento semipermanente
مستوطنة شبه دائمة

3447 semi-rigid foamed plastics
f mousses semi-rigides
e plásticos celulares semirrígidos
لدان رغوية شبه صلبة

3448 sensitive area
f zone sensible
e zona sensible
منطقة حساسة

3449 sensitive species
f espèce sensible
e especie sensible
انواع حساسة

3450 sensitive to pollution
f sensible à la pollution
e sensible a la contaminación
حساس للتلوث

3451 sensitivity scale
f échelle de sensibilité
e escala de sensibilidad
جدول الحساسية

3452 sensitivity to pollution
f polluosensibilité
e sensibilidad a la contaminación
الحساسية للتلوث

3453 sensitivity to toxic substances
f sensibilité aux toxiques
e sensibilidad a los tóxicos
الحساسية للمواد السمية

3454 sensitizing potential
f pouvoir sensibilisant
e capacidad de crear conciencia
قدرة على استثارة الحساسية

3455 separate sewer
f système séparatif
e sistema de alcantarillado separado
شبكة مجارى منفصلة

3456 separated sludge
f boues décantées
e fango decantado
حمأة مفصولة

3457 separators
f séparateurs
e separadores
أجهزة فصل

3458 septic tanks
f fosses septiques
e fosas sépticas
صهاريج تعفين

3459 settleable solids
f matières décantables
e sólidos decantables
مواد صلبة قابلة للترسيب

3460 settlements planning
f planification des établissements humains
e planificaión de los asentamientos humanos
تخطيط المستوطنات

3461 settler
f décanteur
e decantador
حوض الترسيب

3462 settling chamber
f chambre de dépôt
e cámara de asentamiento
حجرة الترسيب

3463 settling tank
f cuve de sédimentation
e estanque de sedimentación
خزان ترسيب

3464 settling time
f temps de sédimentation
e tiempo de sedimentación
زمن الترسب

3465 settling velocity
f vitesse de sédimentation
e velocidad de sedimentación
سرعة الترسب

3466 severely restricted chemical
f substance chimique rigoureusement réglementée
e producto químico objeto de una reglamentación estricta
مادة كيميائية مقيدة بشدة

3467 sewage
f égouts
e aguas residuales
مجارى

3468 sewage disposal
f décharge des eaux usées
e eliminación de aguas residuales
التخلص من المجارى

3469 sewage sludge application
f épandage de boues d'épuration
e esparcimiento de fango cloacal
استعمال حمأة المجارى

3470 sewage treatment plants
f usines de traitement des eaux d'égout
e plantas de tratamiento de aguas
محطات معالجة المجاري

3471 sewage treatment systems
f systèmes d'épuration des eaux usées
e sistemas de tratamiento de aguas residuales
شبكات معالجة المجارى

3472 sewer
f égout
e alcantarilla
أنبوبة مجارى

3473 sewer cleansing sludge
f boues de curage d'égouts
e fango de limpieza de alcantarillas
حمأة تنظيف المجارى

3474 sewerage system
f réseau d'égouts
e sistema de alcantarillado
شبكة المجارى

3475 shallow land burial
f mise en décharge peu profonde
e enterramiento a poca profundidad
دفن ارضي ضحل

3476 shanty towns
f bidonvilles
e barrios de viviendas precarias
مدن أكواخ الصفيح

3477 shares of emissions
f parts des émissions
e partes de las emisiones
حصص الانبعاثات

3478 sharp odour
f odeur âcre
e olor acre
رائحة نفاذة

3479 shedding of leaves
f chute des feuilles
e caída de las hojas
سقوط الاوراق

3480 sheet erosion
f érosion en nappe
e erosión laminar
تآكل طبقي

3481 sheet of water
f plan d'eau
e extensión de agua
طبقة مائية

3482 shelf sea
f mer épicontinentale
e mar epicontinental
بحر جرفي

3483 shellfish
f mollusques et crustacés
e mariscos
أسماك صدفية

3484 shellfishery
f conchyliculture
e marisquería
مصائد الاسماك الصدفية

3485 shelter belts
f plantations-abris
e cinturones de protección
احزمة حماية

3486 shelter crisis
f crise du logement
e crisis de la vivienda
ازمة مساكن

3487 sheltering
f mise à l'abri
e provisión de viviendas
ايواء

3488 shifting
f déplacement
e desplazamiento
تحول

3489 shifting bed
f fond mobile
e fondo fluctuante
مجرى متحول

3490 shifting cultivation
f culture itinérante
e agricultura migratoria
زراعة متنقلة

3491 shipment of hazardous wastes
f expédition de déchets dangereux
e transporte de desechos peligrosos
شحن نفايات خطرة

3492 shoal (1)
f haut fond
e bajo fondo
مرتفع مغمور

3493 shoal (2)
f banc (de poissons)
e banco (de peces)
سرب (اسماك)

3494 shops
f magasins
e almacenes
متاجر

3495 shore meadows
f prairies littorales
e praderas costeras
مروج شاطئية

3496 short-lived waste
f déchets à vie courte
e desechos de vida corta
نفايات قصيرة العمر

3497 short-range weather prediction
f prévision météorologique à courte échéance
e pronóstico meteorológico de corto alcance
تنبؤ قصير المدى بالطقس

3498 short-wave radiation
f rayonnement de courte(s) longueur(s) d'onde
e radiación de onda corta
اشعاع الموجات القصير

3499 shredder truck
f benne déchiqueteuse
e camión triturador
شاحنة تفتيت

3500 shrimp hatchery
f élevage de crevettes
e criadero de camarones
مزرعة اربيان

3501 shrubland
f terres arbustives
e zona arbustiva
أراضي الجنبات

3502 side effects of pharmaceutical drugs
f effets secondaires des médicaments
e efectos secundarios de medicamentos
آثار جانبية للعقاقير الصيدلانية

3503 side reaction
f réaction secondaire
e reacción secundaria
تفاعل جانبي

3504 significant biota
f biotes notables
e biota significativa
حيويات مهمة

3505 significant pollution risk
f risque notable de pollution
e riesgo significativo de contaminación
خطر تلوث كبير

3506 simulation
f simulation
e simulación
محاكاة

3507 single polyvalent vaccine
f vaccin polyvalent unique
e vacuna polivalente única
لقاح وحيد ضد أمراض متعددة

3508 single-cell protein
f protéine monocellulaire
e proteína monocelular
بروتين وحيد الخلية

3509 sink
f puits
e sumidero
بالوعة

3510 sink capacity of an ecosystem
f capacité d'absorption d'un écosystème
e capacidad de absorción de un ecosistema
القدرة الامتصاصية لنظام ايكولوجي

3511 sink factors
f facteurs de dissipation
e factores de disipación
عوامل التبديد

3512 sink of greenhouse gases
f puits des gaz à effet de serre
e sumidero de los gases de efecto invernadero
بالوعات غازات الاحتباس الحراري

3513 sinking
f coulage
e espesamiento y hundimiento
اغراق

3514 site climate
f climat local
e clima local
مناخ محلي

3515 site cultivation
f culture du milieu
e cultivo en el emplazamiento
زراعة موضعية

3516 site directed mutagenesis
f mutagénèse spécifique de site
e mutagénesis específica del emplazamiento
تشوه خلقي بسبب الموقع

3517 site disturbance
f modification du milieu
e modificación del medio
اضراب الموقع

3518 site of generation (of wastes)
f lieu de production (des déchets)
e lugar de generación (de desechos)
موقع توليد (النفايات)

3519 site requirements
f conditions du site
e requisitos del emplazamiento
متطلبات الموقع

3520 site restoration
f remise en état des sites
e restauración de sitios
ترميم الموقع

3521 site surveying
f reconnaissance de la station
e reconocimiento del emplazamiento
مسح الموقع

3522 site value
f valeur d'aliénation
e valor del emplazamiento
قيمة الموقع

3523 sitting of industry
f localisation des industries
e ubicación de industrias
مواقع صناعة

3524 skimmer
f récupérateur
e recuperador
مكشطة

3525 skimming
f écrémage
e recuperación
كشط

3526 skin absorption
f absorption cutanée
e absorción cutánea
امتصاص جلدي

3527 skin disorder
f affection cutanée
e afección cutánea
مرض جلدي

3528 skin exposure
f exposition cutanée
e exposición de la piel
تعرض الجلد

3529 skin test
f test cutané
e prueba cutánea
اختبار جلدي

3530 slag
f laitier
e escoria
خبث

3531 slash burning
f écobuage
e quema de broza y residuos
حرق مخلفات الغابات

3532 slash-and-burn farmer
f paysan pratiquant la culture sur brûlis
e agricultor de corta y quema
مزارع متنقل

3533 slaughterhouse waste
f déchets d'abattoirs
e desechos de mataderos
مخلفات المجازر

3534 slimicides
f mucilagicides
e mucilagicidas
مبيدات العفن المخاطي

3535 slowly developing disasters
f catastrophes à évolution lente
e desastres de evolución lenta
كوارث بطيئة التطور

3536 sludge
f boue(s) résiduaires
e fango residual
حمأة

3537 sludge contact process
f procédé de contact de boues
e procedimiento de contacto de fango
طريقة ملامسة الحمأة

3538 small islands
f petites îles
e pequeñas islas
جزر صغيرة

3539 smarting in eyes
f picotement des yeux
e escozor de los ojos
الم شديد في العينين

3540 smelter dust
f poussières de fours industriels
e polvo de horno de fundición
غبار المصهر

3541 smog
f smog
e smog
ضباب دخاني

3542 smoggy bowl
f dôme de pollution
e caldera de smog
قبة ضباب دخاني

3543 smoke prevention
f prévention des émanations de fumée
e prevención del humo
وقاية من الدخان

3544 smoke stain method
f méthode des taches de fumée
e método de las manchas de humo
طريقة التبقع الدخاني

3545 smokeless fuel
f combustible sans fumée
e combustible sin humo
وقود عديم الدخان

3546 smoking
f tabagisme
e hábito de fumar
تدخين

3547 smooth culture
f culture lisse
e cultivo liso
استنبات منتظم

3548 smuts
f escarbille
e carbonilla
سناج

3549 snap sample
f échantillon ponctuel
e muestra puntual
عينة مأخوذة دون تحديد

3550 snow
f neige
e nieve
ثلج

3551 snow cover
f manteau neigeux
e cubierta de nieve
غطاء ثلجي

3552 snow depth
f épaisseur de neige
e espesor del manto de nieve
عمق الثلج

3553 snow gauge
f nivomètre
e nivómetro
مقياس الثلج

3554 snow melt
f eau de fonte
e deshielo de las nieves
فيضان الثلج الذائب

3555 snow stability
f stabilité du manteau neigeux
e estabilidad de la cubierta de nieve
استقرار الثلج

3556 snow water equivalent
f équivalent en eau
e equivalente en agua
المكافئ المائي للثلج

3557 snow-line
f limite des neiges persistantes
e límite de las nieves permanentes
خط الثلج الدائم

3558 snowdrift
f congère
e nieve acumulada
مجروف ثلجي

3559 soaking
f imbibition
e impregnación
اشباع ؛ تشريب

3560 social indicators
f indicateurs socio-culturels
e indicadores sociales
مؤشرات اجتماعية

3561 social response
f réaction de la collectivité
e reacción social
استجابة اجتماعية

3562 social surveys
f enquêtes sociales
e encuestas sociales
عمليات مسح اجتماعية

3563 societal costs
f coûts sociaux
e costos sociales
تكاليف مجتمعية

3564 socio-economic aspects of human settlements
f aspects socio-économiques des établissements humains
e aspectos socioeconómicos de los asentamientos humanos
الجوانب الاجتماعية الاقتصادية للمستوطنات البشرية

3565 socio-economic factors
f facteurs socio-économiques
e factores socioeconómicos
عوامل اجتماعية اقتصادية

3566 socio-economic impact of biotechnologies
f impact socio-économique des biotechnologies
e impacto socio-económico de las biotecnologías
آثار اجتماعية اقتصادية للتكنولوجيات الحيوية

3567 soft detergents
f détergents biodégradables
e detergentes biodegradables
مواد تنظيف قابلة لتحلل

3568 soil
f sol
e suelo
تربة

3569 soil absorption
f absorption par le sol
e absorción por el suelo
امتصاص التربة

3570 soil air
f air au (voisinage du) sol
e aire del suelo
هواء التربة

3571 soil animal population
f faune terricole
e fauna terrícola
حيوانات التربة

3572 soil capabilities
f capacité du sol
e capacidad productiva de suelos
قدرات التربة

3573 soil conservation
f conservation du sol
e conservación de suelos
صيانة التربة

3574 soil contamination
f contamination du sol
e contaminación de suelos
تلوث التربة

3575 soil degradation
f dégradation du sol
e degradación de suelos
تدهور التربة

3576 soil depletion
f épuisement du sol
e empobrecimiento del suelo
استنفاد التربة

3577 soil erosion
f érosion du sol
e erosión del suelo
تآكل التربة

3578 soil fauna
f faune du sol
e fauna del suelo
حيوانات تعيش في التربة

3579 soil humidity
f humidité du sol
e humedad del suelo
رطوبة التربة

3580 soil material
f matériau constitutif du sol
e composición del suelo
مادة ترابية

3581 soil mechanics
f mécanique des sols
e mecánica de suelos
ميكانيكا التربة

3582 soil microflora
f microflore du sol
e microflora del suelo
النباتات المجهرية في التربة

3583 soil moisture
f teneur en eau du sol
e humedad del suelo
رطوبة التربة

3584 soil profile
f profil pédologique
e perfil del suelo
مقطع عرضي للتربة

3585 soil reclamation
f restauration des sols
e recuperación de suelos
استصلاح التربة

3586 soil research
f recherche pédologique
e investigación edafológica
بحوث التربة

3587 soil salination
f salination du sol
e salificación
تملح التربة

3588 soil scientist
f pédologue
e edafólogo
أخصائي في علوم التربة

3589 soil stability
f stalibité du sol
e estabilidad del suelo
استقرار التربة

3590 soil texture
f texture du sol
e granulometría del suelo
قوام التربة

3591 soil type
f type de sol
e tipo de suelo
نوع التربة

3592 soil water
f eau dans le sol
e agua del suelo
مياه التربة

3593 soil with good tilth
f sol ameubli
e suelo con buena capa de tierra de labranza
تربة جيدة للحراثة

3594 soil-forming process
f processus de pédogénèse
e proceso de formación de los suelos
عملية تكوين التربة

3595 soil-plant relationship
f phytopédologie
e relación suelo-planta
علاقة التربة والنبات

3596 soil-water movement
f ressuyage
e desplazamiento del agua del suelo
حركة مياه التربة

3597 solar constant
f constante solaire
e constante solar
هالة الشمس

3598 solar cosmic particle
f particule cosmique solaire
e partícula cósmica solar
جزيئ كوني شمسي

3599 solar cosmic rays
f rayons cosmiques solaires
e rayos cósmicos solares
اشعة كونية شمسية

3600 solar energy
f énergie solaire
e energía solar
طاقة شمسية

3601 solar heating
f chauffage solaire
e calefacción solar
تدفئة شمسية

3602 solar input
f rayonnement solaire (incident)
e radiación solar incidente
الشعاع الشمسي الساقط

3603 solar outburst
f éruption solaire
e erupción solar
انفجار شمسي

3604 solar output
f rayonnement solaire global
e radiación solar total
اجمالي الاشعاع الشمسي

3605 solar radiation
f rayonnement solaire
e radiación solar
إشعاع شمسي

3606 solar wind
f vent solaire
e viento solar
رياح شمسية

3607 solid waste
f déchets solides
e residuos sólidos
نفايات صلبة

3608 solid waste disposal
f évacuation des déchets solides
e eliminación de residuos sólidos
التخلص من النفايات الصلبة

3609 solid waste management
f gestion des déchets solides
e gestión de los residuos sólidos
ادارة النفايات الصلبة

3610 solid wastefill site
f décharge
e vertedero de residuos sólidos
موقع دفن النفايات الصلبة

3611 soluble solids
f exrait sec soluble
e sólidos solubles
جوامد قابلة للذوبان

3612 solute
f soluté
e soluto
مذاب

3613 solution
f solution
e solución
ذوبان ؛ محلول

3614 solvent extraction
f extraction au solvant
e extracción por solventes
استخلاص بالمذيبات

3615 solvent vapour treatment process
f procédé de traitement des vapeurs du solvant
e procedimiento de tratamiento de vapores de solventes
عملية المعالجة بأبخرة المذيبات

3616 soot
f suie
e hollín
سناج

3617 sooty smoke
f fumée fuligineuse
e humo fuliginoso
دخان سناجي

3618 sorbent
f sorbant
e sorbente
مادة ممتزة

3619 sorbent feed
f charge de sorbant
e carga de sorbente
تزويد بالمواد الممتزة

3620 sorbent surface skimmer
f récupérateur oléophile
e recolector por adsorción
مكشطة سطحية ماصة

3621 sorption
f sorption
e sorbción
امتزاز

3622 sorting of household refuse
f tri des déchets ménagers
e separación de basuras
فرز المخلفات المنزلية

3623 sound attenuation
f atténuation phonique
e atenuación del sonido
توهين الصوت

3624 sound ecological balance
f bon équilibre écologique
e equilibrio ecológico adecuado
توازن ايكولوجي سليم

3625 sound environment
f environnement sain
e medio (ambiente) sano
بيئة سليمة

3626 sound management
f gestion active
e buena gestión
ادارة سليمة

3627 sound reduction
f isolement phonique
e atenuación del sonido
خفض الصوت

3628 sound water use
f utilisation rationnelle de l'eau
e utilización racional del agua
استخدام سليم للمياه

3629 sounder
f sondeur
e sondeador
مسبار

3630 sounding
f sondage
e sondeo
سبر

3631 soundproofing
f insonorisation
e insonorización
عزل الصوت

3632 sour gas
f gaz acide
e gas ácido
غاز حمضي

3633 source area
f région d'origine
e zona de orígen
منطقة المصدر

3634 source gases
f gaze à la source
e gases primarios
غازات المصدر

3635 space transportation
f transport spatial
e transporte espacial
نقل فضائى

3636 spaceborne remote sensing
f télédétection spatiale
e teleobservación desde el espacio
استشعار عن بعد من الفضاء

3637 spatial effective resolution element
f élément de résolution spatiale efficace
e elemento de resolución espacial efectiva
عنصر التحليل الحيزى الفعال

3638 spatial environmental planning
f aménagement de l'espace
e planificación del medio espacial
تخطيط بيئى مكانى

3639 spatial resolution
f limite de résolution spatiale
e resolución espacial
تحليل مكانى

3640 spawning bed
f zone de reproduction
e zona de reproducción
منطقة تكاثر

3641 spawning ground (for fish)
f frayère
e desovadero
مسرأ اسماك

3642 specialized aquatic flora
f flore aquatique adaptée
e flora acuática adaptada
نباتات مائية متخصصة

3643 specialized crop species
f espèce cultivée spécialisée
e especie cultivada adaptada
انواع محاصيل متخصصة

3644 species abundance
f abondance des espèces
e abundancia de la especie
وفرة الانواع

3645 species distribution area
f aire de répartition (géographique) d'une espèce
e zona de distribución geográfica de una especie
منطقة توزع الانواع

3646 species diversity
f diversité des espèces
e diversidad de las especies
تنوع الانواع

3647 species-rich biomes
f biomes riches en espèces
e biomas ricos en especies
مناطق أحيائية غنية بالانواع

3648 specific absorbed dose
f dose absorbée spécifique
e dosis absorbida específica
جرعة ممتصة نوعية

3649 specific absorption rate
f taux d'absorption spécifique
e coeficiente de absorción específica
معدل الامتصاص النوعى

3650 specific discharge
f débit spécifique
e descarga específica
تصريف نوعى

3651 specific response plan
f plan particulier d'intervention
e plan específico de medidas
خطة استجابة معينة

3652 specific weight
f poids spécifique
e peso específico
وزن نوعى

3653 specifications
f caractéristiques
e especificaciones
مواصفات

3654 spectral absorption
f absorption spectrale
e absorción espectral
امتصاص طيفى

3655 spectral band
f bande spectrale
e banda espectral
نطاق طيفى

3656 spectral irradiance
f flux spectral
e irradiancia espectral
اشعاعية طيفية

3657 spectral line
f raie spectrale
e línea espectral
خط طيفى

3658 spectral pattern
f spectrogramme
e patrón espectral
نمط طيفى

3659 spectral range
f étendue spectrale
e gama espectral
نطاق طيفى

3660 spectral region
f région spectrale
e región espectral
منطقة الطيف

3661 spectroscopy
f spectroscopie
e espectroscopia
علم الطيف

3662 spent acid
f acide utilisé
e odaiciv odicل
حمض مستهلك

3663 spent fuel
f combustible épuisé
e combustible agotado
وقود مستهلك

3664 sperm count
f numération des spermatozoïdes
e recuento de espermatozoos
عدد الحيوانات المنوية

3665 spill
f déversement (accidentel)
e derrame
انسكاب

3666 spill incident
f cas de déversement
e caso de derrame
حادث انسكاب

3667 spillage
f écoulements accidentels
e derrames
إراقة

3668 spit
f cordon littoral
e flecha litoral
لسان ساحلى

3669 spoil
f déblai
e escombros
مخلفات الحفر

3670 spoilage
f dégradation
e degradación
اتلاف

3671 spontaneous combustion
f autoinflammation
e combustión espontánea
احتراق تلقائى

3672 sports facilities
f infrastructures sportives
e instalaciones deportivas
مرافق رياضية

3673 spot check
f contrôle par sondage
e control al azar
فحص عشوائى

3674 spray
f atomisation
e vaporización
رذاذ

3675 sprayer-washer
f arroseuse-laveuse
e rociadora-lavadora
رشاشة – غسالة

3676 spread of the desert
f progression du désert
e avance del desierto
امتداد الصحراء

3677 spreading
f épandage
e esparcimiento
انتشار

3678 spring
f mouillère
e manantial
منطقة رطبة

3679 sprout
f rejet
e brote de cepa
برعم

3680 stabilizer
f stabilisateur
e estabilizador
عامل مثبت

3681 stable air
f air stable
e aire estable
هواء مستقر

3682 stack conditions
f paramètres d'émission d'une cheminée
e características de emisión de una chimenea
شروط المداخن

3683 stack effluents
f produits de combustion
e efluentes de chimenea
غازات المداخن

3684 stack gas cleaning
f épuration des gaz brûlés
e depuración de gases de chimenea
تنظيف غاز المداخن

3685 stack solids
f envols
e escapes sólidos de chimenea
جوامد غازات المداخن

3686 stagnant atmospheric conditions
f marais barométrique
e estancamiento atmosférico
أحوال جوية راكدة

3687 stakeholders
f parties prenantes
e interesados directos
أطرف مؤثرة

3688 standard solution
f solution type
e solución tipo
محلول معيارى

3689 standards for building industry
f normes et règles de l'industrie du bâtiment
e normas y códigos de la industria de la construcción
معايير صناعة المبانى

3690 staple crop
f culture de base
e cultivos básicos
محصول اساسى

3691 staple diet
f pâture de base
e forraje básico
غذاء اساسى

3692 staple foods
f aliments de base
e alimentos básicos
أغذية أساسية

3693 state variable
f variable d'état
e variable de estado
متغير الحالة

3694 state-of-the-art technology
f techniques de pointe
e la tecnología más moderna
تكنولوجيا حديثة

3695 stationary combustion source
f source de combustion fixe
e fuente fija de combustión
مصدر احتراق ثابت

3696 stationary source
f source fixe
e fuente fija
مصدر ثابت

3697 status of development
f niveau de développement
e estado de desarrollo
حالة التنمية

3698 steady progression
f évolution régulière
e evolución previsible
تقدم مطرد

3699 steam cleaning
f nettoyage à la vapeur
e limpieza con vapor
تنظيف بالبخار

3700 steam coal
f charbon de chaudière
e carbón de calderas
فحم المراجل (البخارية)

3701 steam distillation
f distillation par entraînement à la vapeur
e destilación en corriente de vapor de agua
تقطير البخار

3702 steaming
f entraînement à la vapeur
e vaporizado
معالجة بالبخار

3703 steel industry
f sidérurgie
e industrias siderúrgicas
صناعة الصلب

3704 step-by-step principle
f principe du pas à pas
e principio “paso a paso”
مبدأ التدرج

3705 step-down schedule
f calendrier d’élimination par étapes
e calendario de reducción por etapas
جدول الخفض التدريجى

3706 sterilant
f stérilisant
e esterilizante
معقم

3707 stock solution
f solution mère
e solución madre
محلول أصلى

3708 stocktaking
f évaluation
e evaluación de la situación
جرد

3709 storage of greenhouse gases
f retenue des gaz à effet de serre
e almacenamiento de gases de efecto invernadero
تخزين غازات الاحتباس الحرارى

3710 storage pit
f fosse de stockage
e fosa de almacenamiento
حفرة خزن

3711 storehouses of biodiversity
f réservoirs d’espèces biologiques
e reservas de especies biológicas
مستودعات التنوع البيولوجي

3712 storm sewer
f collecteur d’eaux pluviales
e alcantarilla de aguas de lluvia
مجارى مياه المطر

3713 storm track
f trajectoire des tempêtes
e trayectoria de la(s) tormenta(s)
مسار العاصفة

3714 storm water basin
f bassin d’orage
e depósito de agua de lluvia
حوض تجميع مياه العواصف

3715 storms
f tempêtes
e tormentas
عواصف

3716 straddling stocks
f groupes d’espèces migratrices
e poblaciones compartidas
أرصدة منتشرة فى أكثر من منطقة

3717 strain
f souche
e capa
سلالة

3718 stream flow
f débit d’un cours d’eau
e caudal
تدفق المجرى المائى

3719 stream measurement
f mesure des courants
e medición de corrientes de agua
قياس مجاري المياه

3720 stress area
f zone critique
e zona crítica
منطقة اجهاد

3721 stressful region
f région agressée
e región afectada
منطقة مجهدة

3722 strip mining
f exploitation à ciel ouvert
e minería a cielo abierto
تعدين مفتوح

3723 strong compacting
f compactage intensif
e compactado a fondo
رص شديد

3724 structural adjustment programs
f programmes des ajustements structuraux
e programa de los reajustes estructurales
برامج التكيف الهيكلى

3725 structures
f structures
e estructuras
إنشاءات

3726 stump
f souche
e tocón
جذعة

3727 stumpage price
f prix sur coupe
e precio de la madera en pie
ثمن الخشب فى أرضه

3728 submarines
f sous-marins
e submarinos
غواصات

3729 substituted hydrocarbons
f hydrocarbures substitués
e hidrocarburos sustituidos
هيدروكربونات بديلة

3730 subsurface containment
f enfouissement technique
e enterramiento
دفن تحت السطح

3731 subsurface zone
f subsurface
e subsuperficie
منطقة جوفية

3732 subterranean water
f eaux souterraines
e aguas subterráneas
مياه تحت السطح

3733 subtropical ecosystems
f écosystèmes subtropicaux
e ecosistemas subtropicales
النظم الايكولوجية شبه المدارية

3734 subtropics
f zone subtropicale
e zonas subtropicales
مناطق شبه مدارية

3735 sudden natural disasters
f catastrophes naturelles soudaines
e desastres naturales repentinos
كوارث طبيعية مفاجئة

3736 sudden warming
f réchauffement soudain
e calentamiento repentino
احترار مفاجئ

3737 suitability for climatic conditions
f adaptation au climat
e adaptabilidad al clima
ملاءمة للظروف المناخية

3738 sullage
f eaux ménagères
e aguas residuales
مياه المجارى المنزلية

3739 sulphates
f sulfates
e sulfatos
كبريتات

3740 sulphur content
f teneur en soufre
e contenido de azufre
نسبة الكبريت

3741 sulphur cycle
f cycle du soufre
e ciclo del azufre
دورة الكبريت

3742 sulphur deposition
f dépôt (s) de soufre
e deposición de azufre
ترسب كبريتى

3743 sulphur-containing substance
f substance soufrée
e sustancia que contiene azufre
مادة تحتوى على الكبريت

3744 sulphuric acid
f acides sulfuriques
e ácido sulfúrico
حامض الكبريتيك

3745 sump
f bassin (de vidange)
e pozo de recolección
حوض تجميع

3746 sunspot
f tache solaire
e mancha solar
بقعة شمسية

3747 superficial microlayer
f film ultrasuperficiel
e microcapa superficial
طبقة مجهرية سطحية

3748 supersalinity
f sursalure
e hipersalinidad
ملوحة شديدة

2749 supportive environment
f milieu favorable
e medio (ambiente) favorable
بيئة ملائمة

3750 suppression of the body's immune system
f immunosuppression
e inmunosupresión
اخماد مناعة الجسم

3751 surcharges on fuel prices
f majoration du prix des combustibles
e recargo del precio de los combustibles
رسوم اضافية على اسعار الوقود

3752 surf line
f ligne de brisement des vagues
e línea de rompientes
خط الزبد

3753 surface active agent
f agent de surface
e agente tensoactivo
عامل ذو فاعلية سطحية

3754 surface albedo
f albédo de la surface
e albedo de superficie
انعكاسية السطح

3755 surface collection agent
f agent repousseur
e agente de contención
عامل تجميع سطحى

3756 surface drainage
f drainage de surface
e drenaje de superficie
صرف سطحى

3757 surface layer
f couche superficielle
e capa superficial
طبقة سطحية

3758 surface ozone
f ozone dans l'environnement
e ozono en el medio ambiente
اوزون السطح

3759 surface run-off
f écoulement de surface
e derrame superficial
جريان سطحي

3760 surface soil
f sol de surface
e suelo de superficie
تربة سطحية

3761 surface temperature
f température en surface
e temperatura de la superficie
حرارة السطح

3762 surface tension
f tension superficielle
e tensión superficial
توتر سطحى

3763 surface water
f eaux de surface
e aguas de superficie
مياه سطحية

3764 surface water balance
f bilan de surface
e balance hídrico de superficie
ميزانية المياه السطحية

3765 surface water hydrology
f hydrologie de surface
e hidrología de superficie
علم المياه السطحية

3766 surface wind
f vent en surface
e viento de superficie
رياح سطحية

3767 surface yield
f apports de surface
e rendimiento de superficie
مورد سطحى

3768 surface-based observing station
f station d'observation basé sur le sol
e estación de observación desde el suelo
محطة مراقبة سطحية

3769 surface-induced process
f processus induit à la surface
e proceso inducido en la superficie
عملية متأثرة بالسطح

3770 surge
f aflux
e aflujo
تموج

3771 survivability (of the organism)
f chances de survie (de l'organisme)
e capacidad de supervivencia
قابلية (كائن) للبقاء

3772 suspended load
f charge en suspension
e carga en suspensión
كمية العوالق

3773 suspended matter
f matières en suspension
e sólidos en suspensión
مواد عالقة

3774 suspended particulate matter
f particules en suspension
e partículas en suspensión
جزئيات عالقة

3775 suspended sediment discharge
f débit solide en suspension
e descarga de sedimentos en suspensión
تصريف الرواسب العالقة

3776 suspended solids
f matières en suspension
e sólidos en suspensión
عوالق صلبة

3777 sustainability of ecological systems
f viabilité des systèmes écologiques
e sostenibilidad de los sistemas ecológicos
استدامة النظم الايكولوجية

3778 sustainability technologies
f technologies de la durabilité
e tecnologías de la sostenibilidad
تكنولوجيات الاستدامة

3779 sustainable
f durable
e sostenible
مستدام

3780 sustainable agriculture and rural development
f développement agricole et rural durable
e desarrollo agrícola y rural sostenible
الزارعة المستدامة والتنمية الريفية

3781 sustainable catch
f prise raisonnable
e capturas sostenibles
الصيد المستدام

3782 sustainable development
f développement durable
e desarrollo sostenible
تنمية مستدامة

3783 sustainable development indicators
f indicateurs d'un développement durable
e indicadores del desarrollo sostenible
مؤشرات التنمية المستدامة

3784 swamp gas
f gaz des marais
e gas de pantanos
غاز المستنقعات

3785 sweep-over
f lécher
e barrer
كسح

3786 sweeper-collector
f balayeuse-ramasseuse
e barredor-recolector
مكنسة – مجمعة

3787 swell-shrink soils
f sols dilatables et rétractables
e suelos dilatables y contráctiles
اتربة منتفشة ومتقلصة

3788 swidden cultivation
f culture sur brûlis
e agricultura de quema
زراعة الأرض بعد حرقها

3789 sylviculture
f sylviculture
e silvicultura
زراعة الاشجار

3790 synecology
f synécologie
e sinecología
الايكولوجيا المتزامنة

3791 synergistic effects of toxic substances
f effets synergiques des produits toxiques
e efectos sinérgicos de sustancias tóxicas
آثار منشطة للمواد السامة

3792 synoptic climatology
f climatologie synoptique
e climatología sinóptica
علم المناخ الاجمالى

3793 synoptic disturbance
f perturbation synoptique
e perturbación sinóptica
اضطراب اجمالى

3794 synoptic meteorology
f météorologie synoptique
e meteorología sinóptica
علم الارصاد الجوية الاجمالية

3795 synoptic observation
f observation synoptique
e observación sinóptica
مراقبة شاملة

3796 synoptic station
f station synoptique
e estación sinóptica
محطة المراقبة الشاملة

3797 synthetic detergents
f détergents synthétiques
e detergentes sintéticos
منظفات اصطناعية

3798 synthetic natural gas
f gaz naturel de synthèse
e gas natural sintético
غاز طبيعى تركيبى

3799 synthetic textile fibers
f fibres textiles synthétiques
e fibras textiles sintéticas
ألياف نسيج اصطناعية

3800 systematic name
f nom systématique
e nombre sistemático
اسم تصنيفى

3801 systemic poison
f toxique à action générale
e tóxico de efecto general
سم ذو مفعول عام

T

3802 tagged
f marqué
e marcado
معلم

3803 tail assay
f teneur résiduelle
e análisis de residuos
اختبار محتوى المترسبات

3804 tail recession
f tarissement non influencé
e descenso del caudal
نضوب طبيعى

3805 tailings
f résidus
e residuos
نفايات التقطير

3806 tangible wealth
f patrimoine corporel
e patrimonio tangible
ثروة حقيقية

3807 tar pit
f trou de bitume
e hoyo de alquitrán
حفرة قار

3808 tar production
f traitement du goudron
e producción y uso del alquitrán
إنتاج القطران

3809 tar sands
f sables asphaltiques
e arenas alquitranadas y esquistos
رمل قطرانى

3810 tar use
f utilisation des goudrons
e uso del alquitrán
استخدام القطران

3811 target species
f espèce ciblée
e especie elegida como objetivo
انواع مستهدفة

3812 target value
f valeur cible
e valor fijado como objetivo
قيمة مستهدفة

3813 tax allowances for anti-pollution investments
f système d'abattement fiscal pour les investissements antipollution
e desgravación fiscal por inversiones contra la contaminación
تخفيضات ضريبية مقابل الاستثمارات فى مكافحة التلوث

3814 tax differentiation
f différenciation des prélèvements fiscaux
e diferenciación de las tasas fiscales
تفاضل ضريبى

3815 **tax incentives**
f mesures d'incitation fiscale
e incentivos fiscales
حوافز ضريبية

3816 **tax relief**
f exonération
e desgravación fiscal
اعفاء ضريبى

3817 **taxa**
f taxons
e taxones
أنواع

3818 **taxonomy**
f taxonomie
e taxonomía
علم التصنيف

3819 **technical backstopping**
f soutien technique
e apoyo técnico
دعم تقنى

3820 **technical grade**
f qualité technique
e calidad técnica
رتبة صناعية

3821 **technological disasters**
f catastrophes technologiques
e desastres tecnológicos
كوارث تكنولوجية

3822 **technological hazards**
f risques technologiques
e peligros tecnológicos
اخطار تكنولوجية

3823 **technological literacy**
f initiation technologique
e conocimientos básicos de tecnología
معرفة تكنولوجية

3824 **technology assessment**
f évaluation technologique
e valoración de la tecnología
تقييم التكنولوجيا

3825 **technology transfer**
f transfert de technologies
e transferencia de tecnologías
نقل التكنولوجيا

3826 **telecommunications**
f télécommunications
e telecomunicaciones
الاتصالات

3827 **telemetry**
f télémesure
e telemetría
قياس عن بعد

3828 **telemorph stage**
f stade télémorphe
e etapa telemórfica
مرحلة التكون المعزول

3829 **temperate ecosystems**
f écosystèmes tempérés
e ecosistemas de zona templada
النظم الايكولوجية للمناطق المعتدلة

3830 temperate forest biosphere
f biosphère en forêts tempérées
e biosfera de bosques de las zonas templadas
المحيط الحيوى للغابات المعتدلة

3831 temperate forests
f forêts des zones tempérées
e bosques de zona templada
غابات معتدلة

3832 temperate woodlands
f régimes boisés des zones tempérées
e regiones arbolades templadas
حراج معتدلة

3833 temperature inversion
f inversion de température
e inversión de temperatura
انقلاب حرارى

3834 temperature-humidity index
f indice de température-humidité
e arutarepmet ed ecidnی
dademuh
الرقم القياسىللحرارة والرطوبة

3835 temporal resolution
f résolution temporelle
e resolución temporal
تحليل زمنى

3836 temporary housing
f logements temporaires
e viviendas temporales
إسكان مؤقت

3837 temporary shelter
f abris d'urgence
e refugios provisionales
مأوى مؤقت

3838 tenant farmer
f fermier
e aparcero
مزارع مستأجر

3839 tentative standard
f norme expérimentale
e norma provisional
معيار مؤقت

3840 teratogen
f agent tératogène
e agente teratogénico
مشوه

3841 terminal leader
f pousse apicale
e crecimiento apical
نامية نهائية

3842 terrain
f terrain
e terreno
تضاريس أرضية

3843 terrestrial biological resources
f ressources biologiques terrestres
e recursos biológicos terrestres
موارد بيولوجية أرضية

3844 terrestrial biota
f biote terrestre
e biota terrestre
حيويات برية

3845 terrestrial ecosystems
f ecosystèmes terrestres
e ecosistemas terrestres
النظم الايكولوجية الأرضية

3846 terrestrial radiation
f rayonnement tellurique
e radiación terrestre
الاشعاع الأرضى

3847 tertiary treatment
f traitement tertiaire
e tratamiento terciario
معالجة ثالثة

3848 test observatory
f observatoire expérimental
e observatorio experimental
مرصد تجريبي

3849 test on workability
f test de fonctionnalité
e prueba de viabilidad
اختبار الصلاحية

3850 test sieve
f tamis de contrôle
e tamiz de prueba
غربال الاختبار

3851 test site
f site de l'essai
e lugar de la prueba
موقع تجارب

3852 thaw behaviour
f comportement au dégel
e comportamiento en el deshielo
سلوك ذوبان الجليد

3853 thematic map
f carte thématique
e mapa temático
خريطة موضوعية

3854 thermal adjustment time
f temps d'adaptation thermique
e tiempo de ajuste térmico
زمن التكيف الحرارى

3855 thermal discharge
f rejets thermiques
e descarga térmica
تفريغ حرارى

3856 thermal efficiency
f rendement thermique
e rendimiento térmico
كفاءة حرارية

3857 thermal expansion
f dilatation thermique
e dilatación térmica
تمدد حرارى

3858 thermal fluid
f fluide thermique
e fluido térmico
سائل حرارى

3859 thermal image
f image thermographique
e imagen térmica
صورة حرارية

3860 thermal input
f apport thermique
e insumo térmico
مدخلات حرارية

3861 thermal insulation
f isolation thermique
e aislamiento térmico
عزل حرارى

3862 thermal load
f charge thermique
e carga térmica
حمل حرارى

3863 thermal losses
f ponts thermiques
e pérdidas térmicas
فقد الحرارة

3864 thermal plume
f panache thermique
e penacho térmico
عمود حرارى

3865 thermal pollution
f pollution thermique
e contaminación térmica
تلوث حرارى

3866 thermal sea power
f énergie thalassothermique
e energía de las mareas
توليد قوى حرارية باستخدام مياه البحر

3867 thermal stratification
f stratification thermique
e termoestratificación
تكون طبقات حرارية

3868 thermocline
f thermocline
e termoclino
المجال الحرارى

3869 thermopause
f thermopause
e termopausa
طبقة الركود الحرارى

3870 thermoset
f thermodurci
e material termofraguable
التصلد بالحرارة

3871 thermosphere
f thermosphère
e termosfera
الغلاف الحرارى

3872 thindown
f dégradation
e degradación
ترقق

3873 thinning of the ozone layer
f diminution de la couche d'ozone
e disminución de la capa de ozono
ترقق طبقة الاوزون

3874 third-order consumers
f consommateurs de troisième ordre
e consumidores terciarios
مستهلكون من الدرجة الثالثة

3875 threatened species
f espèce(s) menacées(s)
e especies en peligro
انواع مهددة

3876 three-membered ring
f cycle triangulaire
e anillo triangular
حلقة ثلاثية

3877 threshold dose
f dose minimale efficace
e dosis umbral
أقل جرعة فعالة

3878 tickle irrigation
f irrigation goutte à goutte
e irrigación por goteo
ري بالتنقيط

3879 tidal drainage system
f réseau d'écoulement des eaux de marée
e sistema de drenaje de aguas mareales
شبكة صرف مياه المد

3880 tidal energy
f énergie marémotrice
e energía de las mareas
طاقة مدية

3881 tidal wave
f raz de marée
e maremoto
موجة مدية

3882 tide gauge
f marégraphe
e escala de marea
مقياس المد

3883 tide station
f observatoire de la marée
e observatorio de marea
محطة مراقبة المد

3884 tide theory
f théorie des marées
e teoría de las mareas
نظرية المد والجزر

3885 tideline
f laisse de haute mer
e marca de marea
خط المد

3886 tides
f marées
e mareas y niveles del mar
مد وجزر

3887 tillage practice
f pratique culturale
e sistemas de labranza
ممارسة الحراثة

3888 timber forest
f forêt d'exploitation
e bosque maderable
غابة اخشاب

3889 timber harvesting
f coupe des forêts d'exploitation
e corta de árboles maderables
قطع الاخشاب

3390 time belt
f fuseau horaire
e huso horario
منطقة توقيت

3891 time scale
f échelle de temps
e escala de tiempos
مقياس زمني

892 **time series**
f chronique
e serie cronológica
سلاسل زمنية

3893 **time-averaged anomalies**
f moyenne temporelle des anomalies
e media temporal de las anomalías
المتوسط الزمنى للظواهر الشاذة

3894 **time-of-wetness principle**
f principe de la durée d'humidité
e principio del período de humedad
مبدأ فترة الرطوبة

3895 **tip burn on leaf**
f brûlure au sommet du limbe
e quemadura del ápice de hoja
احتراق اطراف الورق

3896 **tipping site**
f terrain de décharge
e lugar de descarga
مقلب (نفايات)

3897 **tissue culture**
f culture tissulaire
e cultivo de tejido
استنبات الأنسجة

3898 **titrant**
f solution titrante
e reactivo de valoración
محلول المعايرة

3899 **tobacco**
f tabacs
e tabaco
تبغ

3900 **tolerance**
f tolérance
e tolerancia
تحمل

3901 **top-of-atmosphere fluxes**
f flux au sommet de l'atmosphère
e flujos de la alta atmósfera
تدفقات قمة الغلاف الجوى

3902 **topical action**
f action locale
e acción local
تأثير موضعى

3903 **topsoil**
f sol superficiel
e capa superficial
تربة سطحية

3904 **tort**
f actions en responsabilité
e culpa civil
ضرر

3905 **total airborne fraction**
f fraction atmosphérique totale
e fracción atmosférica total
مجموع الجزيئات المحمولة بالجو

3906 **total cloud cover**
f nébulosité totale
e nubosidad total
مجموع الغطاء السحابى

3907 total column ozone depletion
f appauvrissement de la colonne totale d'ozone
e agotamiento de la columna total de ozono
استنفاد عمود الاوزون الكلى

3908 total deposition
f dépôt total
e deposición total
مجموع الترسبات

3909 total economic value
f valeur économique totale
e valor económico total
القيمة الاقتصادية الكلية

3910 total organic carbon
f carbone organique total
e carbono orgánico total
مجموع الكربون العضوى

3911 total oxidant
f oxydant total
e oxidante total
المؤكسد الكلى

3912 total oxygen demand
f demande totale en oxygène
e demanda total de oxígeno
الحاجة الكلية للاكسجين

3913 total ozone column
f colonne d'ozone total
e columna de ozono total
عمود الاوزون الكلى

3914 total ozone content
f contenu total d'ozone
e contenido de ozono total
المحتوى الكلى الاوزون

3915 total ozone oscillation
f oscillation de l'ozone total
e oscilación del ozono total
تذبذب الاوزون الكلى

3916 total soil water potential
f potentiel hydrique total
e potencial hídrico total del suelo
مجموع الطاقة الكامنة لمياه التربة

3917 total suspended particulates
f particules totales en suspension
e total de partículas en suspensión
مجموع الجزيئات العالقة

3918 tourism
f tourisme
e turismo
سياحة

3919 tourist facilities
f installations touristiques
e instalaciones turísticas
مرافق سياحية

3920 toxic build-up
f accumulation de produits toxiques
e acumulación de sustancias tóxicas
تراكم سمى

3921 toxic substances
f substances toxiques
e sustancias tóxicas
مواد سامة

3922 toxic trace pollutants
f polluants toxiques en trace
e oligocontaminantes tóxicos
ملوثات سمية نزرة

3923 toxic waste
f déchets toxiques
e desechos tóxicos
نفايات سامة

3924 toxicant
f produit toxique
e toxicante
مادة سمية

3925 toxicant monitoring
f toxicovigilance
e control de sustancias tóxicas
رصد المواد السمية

3926 toxicity assessment
f évaluation de la toxicité
e evaluación de la toxicidad
تقييم السمية

3927 toxicity emission factor
f facteur de rejet toxique
e factor de emisión de toxicidad
عامل الانبعاث السمى

3928 toxicity index
f indice de toxicité
e dadicixot ed ecidnي
الرقم القياسى للسمية

3929 toxicity
f toxicité
e toxicidad
السمية

3930 toxicity limit
f seuil de toxicité
e límite de toxicidad
حد السمية

3931 toxicity of pesticides
f toxicité des pesticides
e toxicidad de plaguicidas
سمية مبيدات الآفات

3932 toxicological testing
f tests toxicologiques
e pruebas toxicológicas
اختبارات السمية

3933 toxicology
f toxicologie
e toxicología
علم السموم

3934 toxics
f substances toxiques
e sustancias tóxicas
السميات

3935 toxification of soils
f empoisonnement des sols
e intoxicación de los suelos
تسمم التربة

3936 toxins
f toxines
e toxinas
توكسينات

3937 trace analysis
f analyse des traces
e análisis de trazas
تحليل المواد النزرة

3938 trace element
f élément-trace
e oligoelemento
عنصر نزر

3939 trace elements
f oligo-éléments
e oligoelementos
عناصر نزرة

3940 trace gas
f gaz en traces
e gas en trazas
غاز نزر

3941 trace materials
f substances-traces et résidus
e oligosustancias y residuos
مواد نزرة

3942 tracer
f traceur
e trazador
كاشف

3943 tracer solution
f solution mère
e solución trazadora
محلول كاشف

3944 tracking
f poursuite
e seguimiento
تتبع

3945 tradable emission permit
f permis d'émission négociable
e permiso negociable de emisión
تصريح اطلاق انبعاثات قابلة للتداول

3946 tradable entitlements
f droits négociables
e derechos negociables
استحقاقات قابلة للتداول

3947 tradable permits
f permis négociables
e licencias negociables
تصاريح قابلة للتداول

3948 trade barriers
f restrictions commerciales
e obstáculos al comercio internacional
حواجز تجارية

3949 trade impact on environment
f impacts commerciaux sur l'environnement
e impactos comerciales sobre el medio ambiente
أثر التجارة على البيئة

3950 trade name
f appellation commerciale
e nombre comercial
اسم تجارى

3951 trade winds
f alizés
e alisios
الرياح التجارية

3952 trade-off point
f point d'équilibre
e punto de equilibrio
نقطة التوزان

3953 traditional health care
f médecine traditionnelle
e medicina tradicionalista
رعاية صحية تقليدية

3954 traditional knowledge systems
f modes traditionnels d'acquisition des connaissances
e sistemas tradicionales de adquisición de conocimientos
نظم المعرفة التقليدية

3955 traffic monitoring
f surveillance de la circulation
e vigilancia del tráfico
رصد حركة المرور

3956 traffic noise
f bruits de la circulation
e ruido del tráfico
ضوضاء حركة المرور

3957 training centre
f centre de formation
e centro de formación
مركز تدريب

3958 trait
f caractère
e rasgo
سمة

3959 trans-border disposal
f élimination transfrontière
e eliminación transfronteriza
تصريف عبر الحدود

3960 trans-frontier pollution
f pollution transfrontalière
e contaminación transfronteriza
تلوث عبر الحدود

3961 transboundary air pollution
f pollution atmosphérique transfrontière
e contaminación atmosférica transfronteriza
تلوث جوى عابر للحدود

3962 transboundary movement
f mouvement transfrontière
e movimiento transfronterizo
نقل عبر الحدود

3963 transboundary waters
f eaux transfrontières
e aguas transfronterizas
مياه عابرة للحدود

3964 transfer of a pollutant
f transfert d'un polluant
e traslado de un contaminante
نقل الملوث

3965 transfer of forested land
f conversion de terres forestières
e conversión de tierras boscosas
تحويل الاراضى الحرجية

3966 transformation rate
f taux de conversion
e tasa de transformación
معدل التحول

3967 transgenic animal
f animal transgénique
e animal transgénico
حيوان محور جينيا

3968 transgenic crop
f plante cultivée transgénique
e cultivo de plantas transgénicas
محصول محور جينيا

3969 transgenic plant
f végétal transgénique
e planta transgénica
نبات محور جينيا

3970 transient flow
f écoulement transitoire
e flujo variable
تدفق مؤقت

3971 transit state
f état de transit
e estado de tránsito
دولة العبور

3972 transition phase
f phase de transition
e fase de transición
مرحلة التحول

3973 transitional substances
f substance de transition
e sustancias de transición
مواد تحولية

3974 transpiration
f sudation
e sudación
نتح

3975 transport fuel tax
f taxe sur les carburants
e impuesto sobre los combustibles
ضريبة وقود وسائل النقل

3976 transport of a slick
f déplacement d'une nappe de pétrole
e desplazamiento de una capa de petroléo
انتقال البقعة (النفطية)

3977 transport of hazardous materials
f transport de substances dangereuses
e transporte de materiales peligrosos
نقل المواد الخطرة

3978 transport planning
f planification des transports
e planificación del transporte
تخطيط النقل

3979 transport systems
f systèmes de transport
e sistemas de tranporte
شبكات نقل

3980 transportation
f transports
e transporte
النقل

3981 trap
f collecteur
e eliminador
جهاز تجميع

3982 trap cropping
f culture-piège
e planta trampa
زرع فخى

3983 trapped air
f air retenu
e aire aprisionado
هواء محبوس

3984 trash
f déchets
e basura
قمامة

3985 travel (1)
f propagation
e desplazamiento
انتشار (ملوث)

3986 travel (2)
f voyages
e viajes
سفر

3987 travel time
f temps de propagation
e tiempo de recorrido
زمن الانتشار

3988 travelling standards
f étalons itinérants
e normas móviles
معايير المسافات

3989 treatment plant
f station d'épuration
e planta de tratamiento
منشأة معالجة

3990 tree belt area
f lien forestier
e cinturón arbolado
حزام شجرى

3991 tree biomass
f biomasse ligneuse
e biomasa arbórea
كتلة احيائية شجرية

3992 tree cover
f couvert végétal
e cubierta forestal
غطاء شجرى

3993 tree crops
f cultures arbustives
e arbolados de producción mixta
محاصيل شجرية

3994 tree line
f limite forestière
e límite de la vegetación arbórea
حد نمو الاشجار

3995 tree nurseries
f pépinières
e viveros
مشاتل الأشجار

3996 tree savanna
f savane arborée
e sabana arbórea
سافانا شجرية

3997 tree surgery
f chirurgie arboricole
e cirugía del árbol
جراحة شجرية

3998 treeless
f non boisé
e sin árboles
عديم الاشجار

3999 trees
f arbres
e árboles
أشجار

4000 trial in the open environment
f essai dans l'environnement
e ensayo en condiciones naturales
اختبار فى البيئة الطبيعية

4001 trivial name
f nom vulgaire
e nombre vulgar
اسم دارج

4002 trophic factors
f facteurs trophiques
e factores tróficos
عوامل غذائية

4003 tropical belt
f ceinture intertropicale
e cinturón tropical
حزام مدارى

4004 tropical cyclone
f cyclone tropical
e ciclón tropical
اعصار مدارى

4005 tropical deforestation
f déboisement de forêts tropicales
e deforestación de bosques tropicales
ازالة الغابات المدارية

4006 tropical ecosystems
f écosystèmes tropicaux
e ecosistemas tropicales
النظم الايكولوجية المدارية

4007 tropical forest ecosystems
f écosystèmes des forêts tropicales
e ecosistemas de los bosques tropicales
النظم الايكولوجية للغابات المدارية

4008 tropical forest loss
f pertes en forêts tropicales
e pérdida de bosques tropicales
خسارة الغابات المدارية

4009 tropical forests
f forêts tropicales
e bosques tropicales
غابات مدارية

4010 tropical rain forest
f forêt tropicale ombrophile
e bosque pluvial tropical
غابة مدارية مطيرة

4011 tropical storm
f tempête tropicale
e tempestad tropical
عاصفة مدارية

4012 tropospheric ozone
f ozone troposhérique
e ozono troposférico
أوزون تربوسفيرى

4013 trough
f dépression cyclonique
e zona de bajas presiones
منخفض ضغطى

4014 true specific gravity
f densité absolue
e densidad absoluta
الوزن النوعى الحقيقى

4015 tsunami
f tsunami
e tsunami
امواج سنامية

4016 tunnels
f tunnels
e túneles
أنفاق

4017 turbidimeter
f turbidimètre
e turbidímetro
مقياس العكارة

4018 turbidity
f turbidité
e turbidez
تعكر

4019 turbulent exchange
f échange par turbulence
e intercambio por turbulencia
تبادل دوامى

4020 turbulent flow
f écoulement turbulent
e flujo subterráneo
تدفق داومى

4021 typhoon
f typhon
e tifón
اعصار مدارى

U

4022 umbrella programme
f programme-cadre
e programa global
برنامج جامع

4023 unburnt solids
f imbrûlés solides
e sólidos incombustos
اجسام صلبة غير محروقة

4024 uncompacted tip
f décharge non compactée
e descarga sin compactar
مقلب قمامة غير مضغوطة

4025 uncontrolled clearance
f déboisement non réglementé
e desmonte sin control
قطع الأشجار غير المنظم

4026 uncontrolled emissions
f émission non réglementées
e emisiones sin control
انبعاثات غيرخاضعة لرقابة

4027 uncontrolled tipping
f décharge non contrôlée
e descarga sin control
قلب عشوائى (للنفايات)

4028 under-privileged people
f personnes défavorisées
e personas desfavorecidas
فقراء

4029 undercurrent
f sous-courant
e corriente subsuperficial
تيار تحتى

4030 underflow
f inféro-flux
e corriente subfluvial
تدفق تحتى

4031 underground flow
f écoulement souterrain
e flujo subterráneo
تدفق جوفى

4032 underprediction
f sous-estimation
e subestimación
تنبؤ ناقص

4033 understorey plants
f plantes des sous-bois
e plantas del sotobosque
نباتات أرضية الغابات

4034 uneven-aged forest
f forêt inéquienne
e bosque irregular
غابة ذات اشجارمتفاوتة الاعمار

4035 unleaded
f sans plomb
e sin plomo
خال من الرصاص

4036 unlined landfill
f décharge non étanche
e vertedero sin revestimiento
مدفن غير مبطن

4037 unmanaged ecosystem
f écosystème non exploité
e ecosistema sin explotar
نظام ايكولوجى غير مستغل

4038 unsafe water
f eau insalubre
e agua impura
مياه غير مأمونة

4039 unsanitary conditions
f mauvaises conditions sanitaires
e condiciones insalubres
ظروف غير صحية

4040 unsatisfactory landfill
f décharge défectueuse
e vertedero en malas condiciones
مدفن قمامة لا يستوفى الشروط

4041 unsaturated zone
f zone non saturée
e zona no saturada
منطقة غير مشبعة

4042 unstable atmosphere
f atmosphère instable
e atmósfera inestable
غلاف جوى غير مستقر

4043 unsteady flow
f écoulement variable
e corriente inestable
تدفق غير ثابت

4044 unsteady wind
f vent irrégulier
e viento irregular
ريح غير منتظمه

4045 unsustainable consumption patterns
f modes de consommation non viables
e pautas de consumo no sostenibles
انماط استهلاك غير مستدامة

4046 unsustainable development
f développement non durable
e desarrollo no sostenible
تنمية غير مستدامة

4047 up-slope wind
f vent ascendant
e viento ascendente
رياح صاعدة

4048 upgradable technology
f technique améliorable
e tecnología mejorable
تكنولوجيا ممكنة التطوير

4049 uplands
f hautes terres
e tierras altas
مرتفعات

4050 upper wind
f vent en altitude
e viento de altura
الرياح العليا

4051 upsurge
f prolifération
e proliferación
انتشار

4052 uptake
f absorption
e absorción
امتصاص

4053 upward ozone transfer
f transfert ascendant d'ozone
e transferencia ascendente de ozono
انتقال صاعد للاوزون

4054 upwelling
f remontée d'eau
e corriente ascendente
ارتفاع مياه القاع الى السطح

4055 uranium
f uranium
e uranio
يورانيوم

4056 urban areas
f zones urbaines
e zonas urbanas
مناطق حضرية

4057 urban clean-up
f nettoyage des villes
e aseo de las ciudades
تنظيف المدن

4058 urban climatology
f climatologie urbaine
e climatología urbana
علم المناخ الحضرى

4059 urban decay
f dégradation urbaine
e degradación de zonas urbanas
تدهور حضرى

4060 urban design
f urbanisme
e urbanismo
تصميم حضرى

4061 urban greening
f aménagement des espaces verts
e creación de áreas verdes
تشجيرالمناطق الحضرية

4062 urban land
f terres urbaines
e terrenos urbanos
اراض حضرية

4063 urban management
f urbanisme
e urbanismo
ادارة المناطق الحضرية

4064 urban overcrowding
f surpopulation urbaine
e exceso de población urbana
اكتظاظ المناطق الحضرية

4065 urban poor
f pauvres des villes
e los pobres de las zonas urbanas
فقراء المدن

4066 urban renewal
f rénovation urbaine
e renovación de las zonas urbanas
تجديد حضري

4067 urban slums
f taudis urbains
e barrios de tugurios de las ciudades
احياء حضرية فقيرة

4068 urban stress
f stress urbain
e estrés
إجهاد حضري

4069 urban traffic
f trafic urbain
e tráfico urbano
حركة المرور الحضرى

4070 urban water supply
f alimentation en eau des zones urbaines
e abastecimiento urbano de agua
الإمداد بالمياه الحضرى

4071 use value
f valeur d'usage
e valor de uso
قيمة الاستخدام

4072 used beverage can
f boîte de boisson usagée
e lata de bebidas usada
علبة مشروبات مستعملة

4073 user's fees
f droits d'utilisation
e derechos de uso
رسوم الاستعمال

4074 user benefits
f avantages pour les utilisateurs
e beneficios para los usuarios
منافع للمستعملين

4075 user charges
f redevances pour service rendu
e derechos cobrables a los usuarios
رسوم تقديم الخدمات للمستعملين

4076 user cost
f coût pour l'utilisateur
e costo para el usuario
التكلفة للمستعملين

4077 user-pays-principle
f principe de l'utilisateur payeur
e principio "el usuario paga"
مبدأ "المستعمل يدفع"

4078 uses of forests
f exploitation des forêts
e explotación de los bosques
استغلال الغابات

4079 usual name
f nom commun
e nombre habitual
اسم عادى

4080 utilization of local resources
f utilisation des ressources locales
e uso de recursos locales
استخدام الموارد المحلية

4081 utilization of pesticides
f utilisation de pesticides
e uso de plaguicidas
استخدام مبيدات الآفات

V

4082 vacuum distillation
f distillation sous vide
e destilación al vacío
تقطير فراغى

4083 vacuum insulation
f isolation par le vide
e aislamiento por vacío
عزل بالتفريغ

4084 vagaries of climate
f caprices du climat
e caprichos del clima
تقلبات المناخ

4085 valued ecosystem components
f quantification de l'écosystème
e cuantificación de los componentes del ecosistema
مكونات النظم البيئية المقدرة

4086 vapour degreasing
f dégraissage à la vapeur
e desengrasado por vapor
ازالة الشحوم بالبخار

4087 vapour pressure
f tension de vapeur
e presión de vapor
ضغط البخار

4088 vapour release rate
f débit d'émission de vapeurs
e tasa de emisión de vapores
معدل اطلاق البخار

4089 variable wind
f vent variable
e viento variable
ريح متغيرة

4090 varnishes
f vernis
e pinturas y barnices
ورنيش

4091 vascular plant species
f espèces de plantes vasculaires
e especies de plantas vasculares
انواع النباتات الوعائية

4092 vector breeding site
f site de reproduction des vecteurs
e lugar de reproducción de los vectores
موقع تكاثر ناقلات الأمراض

4093 vector control
f lutte contre les vecteurs
e lucha contra los vectores
مكافحة ناقلات الأمراض

4094 vectors of human diseases
f vecteurs de maladies humaines
e vectores de enfermedades humanas
نواقل الأمراض البشرية

4095 vegetable fuel
f combustible végétal
e combustible vegetal
وقود نباتى

4096 vegetable oil crops
f plantes oléagineuses
e plantas oleaginosas
محاصيل النباتات الزيتية

4097 vegetable oils
f huiles végétales
e aceites vegetales
زيوت نباتية

4098 vegetation belt
f étage de végétation
e piso de vegetación
حزام نباتى

4099 vegetation cover
f couvert végétal
e cubierta vegetal
غطاء خضرى

4100 vegetation limit
f limite de végétation
e límite de la vegetación
حد النمو

4101 vegetation monitoring
f surveillance de la végétation
e vigilancia de la vegetación
رصد النباتات

4102 vegetative propagation
f multiplication végétative
e multiplicación vegetativa
تكاثر خضرى

4103 vehicle inspection
f inspection des véhicules
e inspección de vehículos
تفتيش على المركبات

4104 vent air filter system
f système d'épuration de l'air de ventilation
e sistema de filtrado del aire de ventilación
نظام تصفية هواء التهوية

4105 ventilated gas
f gaz de ventilation
e gas de ventilación
غاز تهوية

4106 verification and compliance
f vérification et observation
e verificación y cumplimiento
التحقق والامتثال

4107 veterinary medicine
f médecine vétérinaire
e medicina veterinaria
طب بيطرى

4108 vibration
f vibrations
e vibraciones
ذبذبة

4109 vicarious use
f utilisation par procuration
e utilización indirecta
استفادة غير مباشرة

4110 violator
f contrevenant
e infractor
مخالف

4111 virology
f virologie
e virología
علم الفيروسات

4112 virtual point source
f source ponctuelle théorique
e fuente puntual hipotética
المصدر الثابت الافتراضى

4113 viruses
f virus
e virus
فيروسات

4114 viscosity breaking
f viscoréduction
e reducción de la viscosidad
تخفيف اللزوجة

4115 volatile organic compounds
f composés organiques volatils
e compuestos orgánicos volátiles
مركبات عضوية متطايرة

4116 volcanic ash cloud
f nuage de cendres volcaniques
e nube de cenizas volcánicas
سحابة رماد بركانى

4117 volcanoes
f volcans
e volcanes
براكين

4118 voluntary agreement
f accord volontaire
e acuerdo voluntario
اتفاق طوعى

4119 vortex
f vortex
e vórtice
دوامة

4120 vortex breakdown
f rupture du tourbillon
e desaparición del vórtice
اختفاء الدوامة

4121 vulnerability
f vulnérabilité
e vulnerabilidad
سرعة التأثر

4122 vulnerability analysis
f analyse de vulnérabilité
e análisis de vulnerabilidad
تحليل مدى القابلية للتأثر

4123 vulnerable to pollution
f sensible à la pollution
e vulnerable a la contaminación
سريع التأثر بالتلوث

W

4124 warm start emissions
f émissions au départ à chaud
e emisiones de arranque en caliente
انبعاثات الانطلاق الدافئ

4125 warming
f échauffement
e calentamiento
احترار

4126 warning
f avertissement;alerte
e alarma;alerta
انذار

4127 warning centre
f centre d'alerte
e centro de alerta
مركز انذار

4128 warning level
f niveau d'alerte
e nivel de alerta
مستوى الانذار

4129 warning period
f période d'alerte
e período de alerta
فترة انذار

4130 warning system
f système d'alerte
e sistema de alerta
نظام انذار

4131 wash(-)out rate
f taux de lavage
e tasa de lavado
معدل الغسل

4132 wash-off
f lessivage (superficiel)
e lavado superficial
غسل سطحى

4133 washing powders
f lessives
e polvos para lavar
مساحيق الغسيل

4134 washings
f eaux de lavage
e líquidos de lavado
مياه الغسيل

4135 waste assimilation capacities
f capacité d'assimilation des déchets
e capacidad de asimilación de desechos
قدرات استيعاب النفايات

4136 waste chemical substances
f déchets de substances chimiques
e desechos químicos
نفايات من المواد الكيميائية

4137 waste conversion techniques
f techniques de conversion des déchets
e técnica de conversión de desechos
تقنيات تحويل النفايات

4138 waste disposal
f décharge des déchets
e eliminación de desechos
تخلص من النفايات

4139 waste disposal in the ground
f décharge des déchets dans le sol
e eliminación de desechos en el suelo
تخلص من النفايات فى الأرض

4140 waste disposal taxes
f taxes sur l'enlèvement des déchets
e impuesto sobre la recogida de desechos
ضرائب على التخلص من النفايات

4141 waste flow
f flux de(s) déchets
e flujo de desechos
تدفق النفايات

4142 waste gas burner
f torchère
e quemador de gas residual
جهاز حرق الغاز المبدد

4143 waste heat
f chaleur résiduelle
e calor residual
حرارة مهدرة

4144 waste lagoon
f bassin de sédimentation
e laguna de sedimentación
بحيرة ترسيب النفايات

4145 waste management
f gestion des déchets
e manejo de desechos
ادارة النفايات

4146 waste methanization
f méthanisation des déchets
e metanización de los desechos
ميثنة النفايات

4147 waste minimization
f minimisation des déchets
e minimización de desechos
خفض النفايات إلى الحد الأدنى

4148 waste plant
f installation de traitement des déchets
e instalación de tratamiento de desechos
منشأة لمعالجة النفايات

4149 waste prevention
f prévention de la production de déchets
e prevención de la producción de desechos
منع انتاج النفايات

4150 waste reclamation
f valorisation des déchets
e aprovechamiento de desechos
استصلاح النفايات

4151 waste recovery
f récupération des déchets
e recuperación de desechos
استعادة النفايات

4152 **waste reduction**
f réduction des déchets
e reducción de desechos
الحد من النفايات

4153 **waste reuse**
f réutilisation des déchets
e reutilización de desechos
إعادة استخدام النفايات

4154 **waste streams**
f flux de déchets
e corriente de desechos
مجارى النفايات السائلة

4155 **waste use**
f utilisation des déchets
e utilización de desperdicios
استخدام النفايات

4156 **waste wood**
f bois de rebut
e desperdicios de madera
نفايات خشبية

4157 **wasteful use**
f gaspillage
e desperdicio
استغلال مبدد

4158 **wastelands**
f terres incultes
e tierras incultas
اراضى بور

4159 **wastes**
f déchets
e desechos
نفايات

4160 **wastewater**
f eaux usées
e aguas residuales
مياه مستعملة

4161 **water balance**
f bilan hydrologique
e balance hídrico
توازن مائى

4162 **water bearing layer**
f aquifère
e capa acuífera
طبقة حاملة للمياه

4163 **water body**
f masse d'eau
e masa de agua
كتلة مائية

4164 **water conservation**
f protection des eaux
e conservación de las aguas
صيانة الماء

4165 **water cycle**
f cycle de l'eau
e ciclo del agua
دورة مائية

4166 **water erosion**
f érosion hydraulique
e erosión hidráulica
تآكل بفعل المياه

4167 **water flow alteration**
f changement du débit des eaux
e modificación del caudal de agua
تعديل تدفق المياه

4168 water harvesting
f récupération de l'eau
e recuperación del agua
جمع المياه

4169 water hole
f point d'eau
e ojo de agua
مورد ماء

4170 water hyacinth
f jacinthe d'eau
e jacinto acuático
بصيل الماء

4171 water level gauge
f limnimètre
e limnímetro
مقياس منسوب الماء

4172 water level recorder
f limnigraphe
e limnígrafo
مسجل منسوب الماء

4173 water loading of the air
f contenu en eau de l'air
e contenido de agua del aire
كمية الرطوبة فى الهواء

4174 water management
f gestion de l'eau
e ordenación de las aguas
ادارة المياه

4175 water pollution
f pollution de l'eau
e contaminación de aguas
تلوث المياه

4176 water pumps
f pompes à eau
e bombas de agua
مضخات مياه

4177 water quality
f qualité de l'eau
e calidad del agua
نوعية المياه

4178 water recirculation
f réutilisation des eaux usées
e recirculación del agua
اعادة تدوير المياه

4179 water resources development
f mise en valeur des ressources en eau
e aprovechamiento de los recursos hídricos
تنمية موارد المياه

4180 water resources assessment
f évaluation des ressources en eau
e evaluación de los recursos hídricos
تقييم الموارد المائية

4181 water resources conservation
f conservation des ressources en eau
e conservación de recursos acuáticos
صيانة موارد المياه

4182 water resources engineering
f génie hydro-économique
e ingeniería de los recursos hídricos
هندسة الموارد المائية

4183 water resources management system
f système de gestion des ressources en eau
e sistema de ordenación de los recursos hídricos
نظام ادارة الموارد المائية

4184 water salination
f salinité de l'eau
e salinificación de aguas
ملوحة المياه

4185 water soluble
f hydrosoluble
e hidrosoluble
قابل للذوبان فى الماء

4186 water sources
f ressources en eau
e fuentes de agua
مصادر المياه

4187 water species
f espèce(s) aquatique(s)
e especies acuáticas
انواع مائية

4188 water stress
f stress hydrique
e estrés por falta de agua
اجهاد مائى

4189 water supply
f fourniture d'eau
e abastecimiento de agua
امدادات المياه

4190 water supply engineering
f génie hydrotechnique
e ingeniería de abastecimiento de agua
هندسة امدادات المياه

4191 water supply system
f alimentation en eau
e red de abastecimiento de agua
شبكة المياه

4192 water system
f système d'eau
e sistema hidrológico
نظام مائى

4193 water table
f surface d'une nappe
e capa freática
مستوى المياه الجوفية

4194 water transportation
f transports par voies navigables
e transporte acuático
نقل مائى

4195 water treatment
f traitement de l'eau
e tratamiento del agua
معالجة المياه

4196 water vapour continuum
f continuum de vapeur d'eau
e continuo de vapor de agua
تواصلية سلسلة بخار الماء

4197 water wells
f puits
e pozos de agua
آبار مياه

4198 water-associated vector
f vecteur associé à l'eau
e vector vinculado con el agua
ناقل للامراض ذو صلة بالماء

4199 water-based disease
f maladie d'origine aquatique
e enfermedad de origen acuático
مرض مائى المنشأ

4200 water-carrying capacity
f capacité de débit
e caudal máximo
قدرة على حمل المياه

4201 water-holding capacity
f capacité de rétention de l'eau
e capacidad de retención del agua
قدرة على الاحتفاظ بالماء

4202 water-inhabiting insect
f insecte aquatique
e insecto acuático
حشرة مائية

4203 water-oil emulsion
f émulsion d'huile et d'eau
e emulsión de agua y petróleo
مستحلب نفطى – مائى

4204 water-related hazard
f risque lié à l'eau
e peligro vinculado con el agua
خطر ذو صلة بالماء

4205 water-salt balance
f équilibre hydrosalin
e equilibrio hidrosalino
توازن الماء والملح

4206 water-saving device
f dispositif permettant d'économiser l'eau
e dispositivo para ahorrar agua
جهاز مقتصد فى استهلاك الماء

4207 water-table decline
f baisse de la nappe phréatique
e descenso de la capa freática
انخفاض منسوب المياه الجوفية

4208 water-table gradient
f pente de la surface de saturation
e pendiente freática
تدرج منسوب المياه الجوفية

4209 water-washed disease
f maladie due au manque d'hygiène
e enfermedad causada por condiciones insalubres
مرض قلة الاغتسال

4210 watercourse
f cours d'eau
e curso de agua
مجرى مائى

4211 waterfowl
f gibier d'eau
e aves acuáticas
طير مائي

4212 waterlogged lands
f terres imprégnées d'eau
e tierras anegadas
أراض غدقة

4213 waterlogging
f engorgement (du sol) par l'eau
e sobresaturación
التغدق (التشبع بالماء)

4214 watershed management
f aménagement des bassins versants
e ordenación de cuencas hidrográficas
إدارة مستجمع مياه

4215 waterside development
f mise en valeur des rives
e desarrollo de riberas
تنمية الساحل

4216 wave and tide theory
f théorie houlomarélogique
e teoría de marejadas y mareas
نظرية الامواج والمد والجزر

4217 wave energy
f énergie de la houle
e energía de las olas
طاقة موجية

4218 wave theory
f théorie des vagues
e teoría de las olas
النظرية الموجية

4219 wavelength region
f zone de longueur d'onde
e banda de longitud de onda
نطاق طول الموجة

4220 weather chart
f carte météorologique
e carta meteorológica
خريطة الطقس

4221 weather modification
f modifications météorologiques
e modificaciones del tiempo
تغير الطقس

4222 weather monitoring
f surveillance météorologique
e vigilancia meteorológica
رصد الطقس

4223 weather pattern
f caractères du temps
e pauta meteorológica
نمط الطقس

4224 weather prediction
f prévisions météorologiques
e pronósticos meteorológicos
تنبؤ بالطقس

4225 weather resistance
f résistance aux agents atmosphériques
e resistencia a la intemperie
مقاومة عوامل الطقس

4226 weather ship
f navire météorologique
e buque meteorológico
سفينة ارصاد جوية

4227 weed clearing
f faucardage
e desherba
ازالة الاعشاب

4228 weeds
f mauvaises herbes
e malezas
أعشاب

4229 weedy river
f rivière herbeuse
e río invadido por hierbas
نهر معشوشب

4230 welding dust
f poussière de soudure
e polvo de soldadura
غبار اللحام

4231 welfare economics
f économie de bien-être
e economía del bienestar
اقتصاد الرفاهية

4232 westerlies
f vents d'ouest
e vientos del oeste
رياح غربية

4233 wet scavenging
f nettoyage par voie humide
e limpieza por vía húmeda
تنظيف بالمياه

4234 wet screening
f filtrage par voie humide
e tamizado por via húmeda
تصفية رطبة

4235 wet deposition
f dépôt(s) humide(s)
e depósitos húmedos
ترسبات رطبة

4236 wet lime scrubbing
f lavage aqueux
e lavado con cal apagada
غسل بالجير

4237 wet meadows
f prairies humides
e praderas húmedas
مروج رطبة

4238 wet scrubbing
f lavage aqueux
e lavado con agua
تنظيف بالماء

4239 wet test
f essai par voie humide
e ensayo o prueba por vía húmeda
اختبار رطب

4240 wet tropical zone
f zone tropicale humide
e zona tropical húmeda
منطقة مدارية رطبة

4241 wetlands
f terres humides
e zonas pantanosas
اراض رطبة

4242 wetlands ecosystems
f écosystèmes des terres humides
e ecosistemas de ciénagas
النظم الايكولوجية للأراضي الرطبة

4243 wetted
f mouillé
e mojado
مرطب

4244 wide area network
f réseau étendu
e red de área extensa
شبكة منطقة واسعة

4245 wild foods
f nourritures sauvages
e alimentos silvestres
اغذية برية

4246 wild natural resources
f ressources naturelles sauvages
e recursos naturales silvestres
موارد طبيعية برية

4247 wild resources
f ressources à l'état sauvage
e recursos en estado natural
موارد برية

4248 wild-type virus
f virus sauvage
e virus salvaje
فيروس جامح

4249 wilderness
f état sauvage
e zonas en estado natural
الحياة البرية

4250 wildfires
f feux de friches
e incendios de bosques
حرائق الغابات الهائلة

4251 wildlands
f espaces naturels
e zonas silvestres
البرارى

4252 wildlife
f espèces sauvages
e fauna y flora silvestres
الاحياء البرية

4253 wildlife conservation
f conservation de la faune
e conservación de la fauna
صيانة الأحياء البرية

4254 wildlife habitats
f habitats sauvages
e hábitat de la vida silvestre
موائل الأحياء البرية

4255 wildlife management
f gestion de la faune sauvage
e ordenación de la fauna
ادارة الاحياء البرية

4256 wildlife population statistics
f statistiques de la vie sauvage
e estadísticas de la vida silvestre
إحصائيات عشائر الأحياء البرية

4257 wildlife sanctuary
f aire protégée pour la flore et la faune sauvages
e refugio de especies silvestres
مأوى احياء برية

4258 wildlife trade
f commerce des espèces sauvages
e comercio de especies silvestres
الإتجار فى الاحياء البرية

4259 wilting point
f point de flétrissement permanent
e punto de marchitamiento
نقطة الذبول

4260 wind energy
f énergie éolienne
e energía eólica
طاقة الرياح

4261 wind erosion
f érosion éolienne
e erosión eólica
تآكل بفعل الريح

4262 wind farming
f agriculture éolienne
e aprovechamiento de la energía eólica
زراعة تستغل طاقة الريح

4263 wind field
f champ de vent
e campo de viento
مجال الرياح

4264 wind loading
f charge exercée par le vent
e carga causada por el viento
تحميل الرياح

4265 wind model
f modèle de vent
e modelo de viento
نموذج رياح

4266 wind scale
f échelle de beaufort
e escala anemométrica
مقياس الرياح

4267 wind shear
f cisaillement du vent
e cizalladura del viento
قص الرياح

4268 wind strength
f vitesse des vents
e fuerza del viento
قوة الرياح

4269 wind stress
f tension du vent
e presión (dinámica) del viento
جهد الرياح

4270 wind system
f système de vents
e sistema de vientos
نظام الرياح

4271 wind tunnel
f tunnel aérodynamique
e túnel aerodinámico
نفق ريحى

4272 windborne
f éolien
e transportado por el viento
محمول بالرياح

4273 windbreak
f coupe-vent
e paravientos
مصد الريح

4274 windmill
f moulin à vent
e molino de viento
طاحونة هوائية

4275 winds
f vents
e vientos
رياح

4276 windstorm
f vents violents
e vendaval
عاصفة رياح

4277 windward
f au vent
e barlovento
بتجاه الريح

4278 wintering ground
f zone d'hivernage
e zona de hibernación
مشتى

4279 withering point
f point de flétrissement
e punto de marchitamiento
نقطة الذبول

4280 women status
f condition sociale de la femme
e condición social de la mujer
وضع المرأة

4281 wood fibres
f fibres de bois
e fibras de madera
الياف خشبية

4282 wood harvesting
f abattage
e corta
قطع الاخشاب

4283 wood preservation
f conservation du bois
e preservación de maderas
حفظ الأخشاب

4284 wood products
f produits du bois
e productos de madera
منتجات خشبية

4285 wood waste
f déchets de bois
e residuo de maderas
نفايات الخشب

4286 wood-based indstries
f industries du bois
e industrias madereras
صناعة خشبية

4287 wood-consuming insect
f insecte xylophage
e insecto xilófago
حشرة مستهلكة للخشب

4288 wooded lands
f espaces boisés
e tierras boscosas
اراض حرجية

4289 wooded meadows
f pré-bois
e praderas con arboledas
مروج حرجية

4290 woodfuel plantation
f plantation d'essences de bois de feu
e plantación de especies para leña
مزرعة وقود خشبى

4291 **woodland ecosystems**
f écosystèmes des régions boisées
e ecosistemas de las zonas arboladas
النظم الايكولوجية للحراج

4292 **woodlands**
f terres boisées
e tierras arboladas
حراج

4293 **woodlot**
f parcelle boisée
e arboleda
منطقة شجرية

4294 **woody plant**
f végétal ligneux
e planta leñosa
نبات خشبى

4295 **woody species**
f essences ligneuses
e especie leñosa
انواع خشبية

4296 **woody tissue**
f tissu ligneux
e tejido leñoso
انسجة خشبية

4297 **working environment**
f environnement du lieu de travail
e ambiente del trabajo
بيئة العمل

4298 **workplace air**
f atmosphère des lieux de travail
e aire del lugar de trabajo
هواء مكان العمل

4299 **wrack**
f varech
e sargazo
حطام

4300 **wretched poverty**
f pauvreté totale
e pobreza total
فقر مفرط

Y

4301 yard waste
f déchets de jardin
e desechos de jardín
نفايات الحدائق المنزلية

4302 yeasts
f levures
e levaduras
خمائر

Z

4303 zero-waste strategy
f stratégie du zéro déchet
e estrategia de cero desechos
استراتيجية اللانفايات

4304 zonal ozone transport
f transport zonal de l'ozone
e transporte zonal del ozono
نقل الاوزون النطاقى

4305 zonal soil
f sol zonal
e suelo zonal
تربة ناضجة

4306 zonation
f zonation
e zonación
تقسيم الى مناطق

4307 zoning
f zonage
e zonificación
تقسيم الى مناطق

4308 zoocenosis
f zoocénose
e zoocenosis
مجموعة حيوانية

4309 zoological gardens
f jardins zoologiques
e jardines zoológicos
حدائق الحيوان

4310 zoology
f zoologie
e zoología
علم الحيوان

4311 zoomass
f zomasse
e zoomasa
كتلة احيائية حيوانية

4312 zooplankton
f zooplancton
e zooplancton
بلانكتون حيوانى

Français

Español

عربي